Introduction to Reconfigurable Supercomputing

Synthesis Lectures on Computer Architecture

Editor
Mark D. Hill, *University of Wisconsin, Madison*

Introduction to Reconfigurable Supercomputing

Marco Lanzagorta, Stephen Bique, Robert Rosenberg

ISBN: 978-3-031-00598-5   paperback
ISBN: 978-3-031-01726-1   ebook

DOI 10.1007/978-3-031-01726-1

A Publication in the Springer series
SYNTHESIS LECTURES ON COMPUTER ARCHITECTURE

Lecture #9
Series Editor: Mark D. Hill, *University of Wisconsin, Madison*
Series ISSN
Synthesis Lectures on Computer Architecture
Print 1935-3235   Electronic 1935-3243

Introduction to Reconfigurable Supercomputing

Marco Lanzagorta
ITT Corporation

Stephen Bique
Naval Research Laboratory

Robert Rosenberg
Naval Research Laboratory

SYNTHESIS LECTURES ON COMPUTER ARCHITECTURE #9

ABSTRACT

This book covers technologies, applications, tools, languages, procedures, advantages, and disadvantages of reconfigurable supercomputing using Field Programmable Gate Arrays (FPGAs). The target audience is the community of users of High Performance Computers (HPC) who may benefit from porting their applications into a reconfigurable environment. As such, this book is intended to guide the HPC user through the many algorithmic considerations, hardware alternatives, usability issues, programming languages, and design tools that need to be understood before embarking on the creation of reconfigurable parallel codes. We hope to show that FPGA acceleration, based on the exploitation of the data parallelism, pipelining and concurrency remains promising in view of the diminishing improvements in traditional processor and system design.

KEYWORDS

Cray XD1, FPGA, DSPLogic, Handel-C, High Performance Computing (HPC), Mitrion-C, Reconfigurable Computing, VHDL, Virtex II Pro, Virtex-4, Xilinx

(ML)

*To Granddaddy Fernando, Aunt Lucy, Uncle Jorge,
and Baby Oliver. In loving memory, and with eternal gratitude.*

(SB)

To Anna-Maria, Sylvester, Linda, and Abigail

(RR)

To Kay and Sarah

Contents

Preface

This book covers technologies, applications, tools, languages, procedures, advantages, and disadvantages of reconfigurable supercomputing using Field Programmable Gate Arrays (FPGAs). The target audience is the community of users of High Performance Computers (HPC) who may benefit from porting their applications into a reconfigurable environment. As such, this book is intended to guide the HPC user through the many algorithmic considerations, hardware alternatives, usability issues, programming languages, and design tools that need to be understood before embarking on the creation of reconfigurable parallel codes.

However, this book is not intended to teach how to use and program FPGAs. In particular, this document does not replace the specific documentation provided by the vendors for any of the FPGA programming languages and tools. Before attempting to program a device, the reader is encouraged to read and study the latest documentation for each of the chosen tools.

The first chapter begins with a brief discussion of the technology behind FPGA devices. This discussion covers the basic architecture of an FPGA in terms of logic blocks and interconnects, as well as the programming of these devices. Chapter 2 explains how FPGAs are embedded in supercomputing architectures and how to create reconfigurable parallel codes. The focus of the second chapter is the architecture and functionality of the Cray XD1 reconfigurable supercomputer. Chapter 3 presents a series of algorithmic considerations that are necessary to determine when a computational code is a good candidate for FPGA acceleration. Chapter 4 offers an overview of some of the most widely used FPGA programming languages: VHDL, DSPLogic, Mitrion-C, and Handel-C. Chapter 5 provides a detailed study of how to use reconfigurable supercomputing techniques to accelerate a variety of data sorting algorithms. Finally, in Chapter 6, we discuss recent alternative technologies and summarize our experience with FPGAs.

Marco Lanzagorta, Stephen Bique, and Robert Rosenberg
October 2009

Acknowledgments

The authors acknowledge the support received by Dr. Jeanie Osburn, the NRL-DC HPC Center Manager, and Director Barth Root of the Center for Computational Science, both at the US Naval Research Laboratory. Wendell Anderson (Naval Research Laboratory), and Olaf Storaalsli (Oak Ridge National Laboratory) provided support and numerous valuable suggestions. The authors also appreciate the interesting and insightful discussions with Jace Mogill (formerly with Mitrionics) and Mike Babst (DSPLogic). We additionally acknowledge the encouragement and support received from both editor Dr. Mark Hill and publisher Mike Morgan. Finally, much of the content of this book is based on work entirely performed on the Cray XD1 system at NRL-DC under the auspices of the U. S. Department of Defense (DoD) High Performance Computer Modernization Program (HPCMP).

Marco Lanzagorta, Stephen Bique, and Robert Rosenberg
October 2009

Introduction

Recent years have witnessed reconfigurable supercomputing spearheading a radically new area in high performance computer design. Reconfigurable supercomputers typically consist of a parallel architecture where each computational node is made of traditional ASIC CPUs, working alongside a Field Programmable Gate Array (FPGA) in a master-slave fashion. Reconfigurable supercomputers appear to be of relevance to a variety of codes of interest to the scientific, financial, and defense communities.

The distinctive feature of an FPGA is that its hardware configuration takes place *after* the manufacturing process. This means that the user is no longer limited to a fixed, unchangeable, and predetermined set of hardware functions. That is, an FPGA can create a temporary hardware unit that specifically conducts the types of operations needed by an application. For improved performance, the FPGA is exclusively used to accelerate portions of the code that exhibit a large degree of data parallelism, instruction concurrency, and pipelining.

Therefore, using FPGA technology, supercomputers are able to modify their logical circuits at runtime. As a consequence, these machines can, in principle, work with an innumerable number of circuits designs that are best suited to the specific application of interest. In this regard, it is important to mention that reconfigurable supercomputers have been benchmarked to provide dramatic increases in computational performance for selected examples as well as increased power efficiency [34]. Specifically, this technology has been proved to be more effective for certain applications in the areas of cryptology, signal analysis, image processing, searching and sorting, and bioinformatics.

The co-founders of Xilinx, Ross Freeman, and Bernard Vonderschmitt, invented the FPGA in 1985. Subsequently, Steve Casselman of the US Naval Surface Warfare Center created the first design of a reconfigurable computer implementing 600,000 FPGAs during the late 1980s [43]. Hence, reconfigurable supercomputing using FPGAs is a relatively new technology.

Reconfigurable supercomputing poses many challenges. In particular, parallel programming is not trivial even without FPGAs. Because of the reconfigurable nature of FPGAs, programmers must deal with a variety of hardware and software considerations while designing efficient reconfigurable codes. To date, there is no compiling option that automatically enables FPGA acceleration.

In general, the goal for the software designer of reconfigurable applications is to exploit the specific benefits, flexibility, and efficiency of this technology while avoiding its many known shortcomings. Integrating FPGAs in system design in a way that reduces the overhead of using them will significantly advance the technology. Features such as shared memory are needed to reduce communication costs. Either the time needed to load logic needs to be reduced or the possibility of switching FPGAs needs to be added to make runtime reconfiguration useful.

Furthermore, all parallel codes will not benefit from FPGA acceleration. Programmers must carefully analyze their applications to determine if they are good candidates for FPGA acceleration. Typically, code implemented on an FPGA is a compute-intensive kernel, which is a bottleneck of an application. Testing involves rewriting major portions of code for both the host processor and the co-processor. By any means, reconfigurable software development is a complex and time-consuming task for the neophyte FPGA programmer.

CHAPTER 1

FPGA Technology

The FPGA is a digital integrated circuit that can be configured and programmed by the user to perform a variety of computational tasks [19]. These units work with integers or floating numbers, perform several types of arithmetic and logical operations, access and reference memory locations, and carry out sophisticated control structures involving loops and conditionals. Arguably, the way in which FPGAs merge hardware and software is what makes them unique. Consequently, the design of a reconfigurable algorithm will involve a variety of software and hardware considerations. Indeed, the programmer needs to understand not only the algorithmic structure of the problem to be solved but also the important details about the configuration of the hardware for optimal performance.

1.1 ASIC VS. FPGA

To better appreciate the differences between programming traditional devices and reconfigurable units, let us think about the differences between traditional hardware and FPGAs. With traditional Application Specific Integrated Circuits (ASIC), we can design, compile, and run an arbitrary computer program to perform an arbitrary computation. However, the translation to machine executable controls, made by the compiler, is restricted to the existing operations available inside the CPU.

On the other hand, the hardware configuration of an FPGA takes place after the manufacturing process, which means that programming is no longer limited to a fixed, unchangeable, and predetermined set of hardware functions. Hence, an FPGA becomes a temporary hardware unit that specifically conducts the types of operations needed by an application. This capability justifies the "field programmable" terms in FPGA.

Indeed, a traditional ASIC CPU has a fixed and determined number of integer arithmetic units and floating point units, which may be in use or idle at any moment in time, depending on the specific mathematical or logical operation being carried out. Clearly, ASIC CPUs are designed for flexibility to tackle a wide variety of computational problems. Yet, these CPUs present many challenges to efficiently utilize their hardware resources.

In particular, the architecture of the ASIC CPU may cause a series of delays due to hardware conflicts. In other words, an operation may not be carried out until a hardware unit becomes available. However, an FPGA can, in principle, be programmed in such a way that all of its configured hardware resources will be available at all times. Furthermore, FPGAs offer the potential of an extremely high level of utilization of hardware resources.

For example, let us suppose that we have an application that extensively uses integer additions and not much else. If we use a traditional ASIC CPU, then our application is basically restricted to the number of integer arithmetic adders available while multipliers and other hardware functions remain idle for most of the computational time. On the other hand, if we program this application using an FPGA, then we can configure the device to consist of mainly integer arithmetic adders. Then, the entire hardware resources available inside an FPGA will be busy most of the time.

1.2   PHYSICAL ARCHITECTURE

In a nutshell, an FPGA is merely a semiconductor device made of a grid of programmable logic components (*logic blocks*) tied together by means of *programmable interconnects* [10]. The logic blocks are able to perform combinational logic and flip flops for sequential logic operations. About 10% of the FPGA is made of logic blocks.

The other 90% of the FPGA is made of the programmable interconnects, which form a routing architecture that provides arbitrary wiring between the logic blocks. A simplified abstract view of an FPGA is shown in Figure 1.1. The square boxes represent the logic blocks and the lines correspond to the interconnect structure.

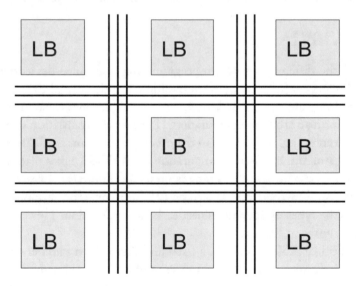

Figure 1.1: Abstract view of an FPGA. The square boxes represent the logic blocks (LB) and the straight lines correspond to the interconnect structure.

1.3   LOGIC BLOCKS

As mentioned before, the logic blocks inside an FPGA can be programmed to perform a variety of functions. In particular, they can be used to carry out the following functions:

• A standard set of logical gates such as AND and XOR.

- Complex combinational mathematical functions.

- High level arithmetic and control structures such as integer multipliers, counters, decoders, and many more.

The great flexibility of the logic blocks is accomplished through the extensive use of *Look Up Tables* (LUTs). The structure of these LUTs is replicated multiple times over the area of the FPGA. As such, LUTs are the basic building block of the logical structure of an FPGA.

As their name suggests, LUTs are simply hardware implementations of logical truth tables. Furthermore, truth tables can be used to represent arbitrary Boolean functions. As such, the FPGAs can be programmed to carry out sophisticated algorithms by using truth tables that represent the Boolean functions relevant to the program.

a	b	c	a XOR b XOR c
0	0	0	0
0	0	1	1
0	1	0	1
0	1	1	0
1	0	0	1
1	0	1	0
1	1	0	0
1	1	1	1

Table 1.1: Truth Table for a 3-bit XOR

As an example, let us consider the case of a very simple Boolean function, the 3-bit XOR. This function is represented by f(a,b,c) = a XOR b XOR c, whose output depends on the eight possible inputs as shown in a truth table in Table 1.1.

An abstract diagrammatic representation of the hardware implementation of the 3-bit XOR as an LUT can be seen in Figure 1.2. The crossed vertical line in the lower part of the figure denotes an input of 3 bits. Depending on the value of the 3-bit input, the LUT selects the value of the function from the column at the left-hand side of the figure. That is, if the input is 000, then the LUT outputs 0, the number of the column with an address 000; if the input is 001, then the LUT outputs 1, the number of the column with an address 001; and so on.

Those familiar with logical circuits will appreciate that the LUT is being implemented as a multiplexer unit. That is, a multiplexor is a device that selects one of many input signals and sends it as a single output signal. In general, a multiplexer with 2^n available inputs requires a selector made of n bits. The value of the selector is the address of the input being sent to the output of the multiplexer.

Clearly, an arbitrary Boolean function can be computed with an LUT. However, the LUT has to be large enough to fully describe all of the possible inputs of the Boolean function. A simple logical gate such as an XOR may allow four inputs described by a 2-bit input signal, but a more

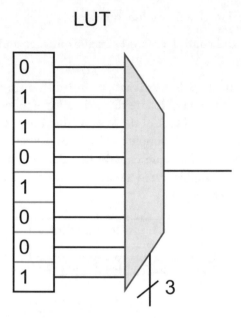

Figure 1.2: Diagrammatic representation of the hardware implementation of the 3-bit XOR as an LUT.

sophisticated Boolean function operating on 32-bit integers will permit 2^{32} inputs described by a 32-bit input signal. In general, an n-LUT is an LUT that selects one out of 2^n input signals using an n-bit selector signal.

In the context of FPGA design, there are advantages and disadvantages of using large or small LUTs. For instance, FPGAs with large LUTs permit the easy description of complex logical operations and require less wiring in the interconnect. On the other hand, FPGAs with large LUTs may yield sluggish performance because they require larger multiplexers. Also, they may end up wasting valuable hardware resources if simple Boolean functions are programmed inside a large LUT.

If the FPGA uses small LUTs, then the programming of a Boolean function may require a large number of logical blocks and extensive wiring in the interconnect. Furthermore, the extensive wiring between blocks may be a cause of delay, which may lead to slower computational performance.

Therefore, LUTs inside an FPGA cannot be too big or too small. To date, it appears that the most optimal and efficient trade-off between the size of the LUT and the required wiring in the interconnect is a 4-LUT. Nevertheless, recent FPGA architectures such as the Virtex-5 use 6-LUTs. Furthermore, optimal operational design of FPGAs often groups more than a single 4-LUT. In general, an *FPGA slice* is usually made of a pair of 4-LUTs or 6-LUTs.

It is important to note that the abstract representation given in Figure 1.2 is not completely accurate. The multiplexer does not store state information within the logic block. Hence, it is impossible to use it to perform any type of sequential or state-holding logical operations. The solution to this deficiency is easy. As shown in Figure 1.3, the multiplexer requires a single-bit storage element implemented with a D flip-flop. This flip-flop simply stores the result of the 4-LUT until the next signal.

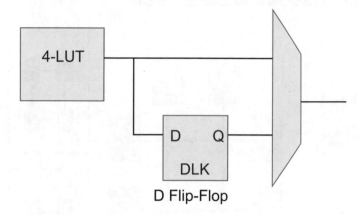

Figure 1.3: Logic block design, including a D Flip-Flop unit for single-bit storage.

1.4   THE INTERCONNECT

As previously discussed, the interconnect is configured to properly route all of the signals between the logic blocks inside an FPGA. In general, large computations are broken into simpler operations that encompass several logic blocks. Then, the interconnect is used to gather the results of the individual computations that were performed by all the LUTs involved in the computation. An FPGA can perform arbitrary computations as long as the logic blocks and the interconnect are large enough to fully describe all of the required logical operations. There are several ways to structure the interconnect:

- Nearest Neighbor.

- Segmented.

- Hierarchical.

The simplest of all are the *nearest neighbor* structures. As their name suggests, nearest neighbor interconnect structures only connect logic blocks to the nearest logic blocks. Unfortunately, this structure often leads to severe delays in passing information from one side of the FPGA to another (a delay that scales linearly with the distance required to be traversed) and also incurs a wide variety of connectivity issues (signals passing through a logic block should not interfere with new signals being produced inside the logic block being traversed).

Segmented structures are a more sophisticated way to implement the interconnect. In addition to LUTS, these structures rely on connection blocks and switch boxes to provide increased routing flexibility. Finally, *hierarchical structures* group together logic blocks in a segmented structure accommodated in a hierarchical manner. In practice, most modern FPGA architectures implement a variant of segmented or hierarchical interconnect structures.

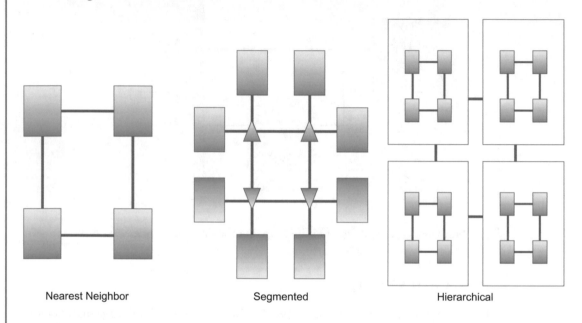

| Nearest Neighbor | Segmented | Hierarchical |

Figure 1.4: Nearest Neighbor, Segmented and Hierarchical interconnect structures. Squares represent logic blocks, lines the interconnect, and the triangles are connection and switching boxes.

1.5 MEMORY AND I/O

While the basic components of an FPGA are the logic blocks and the interconnect, most commercial FPGAs have a slightly more sophisticated architecture. New components need to be added to the architecture in order to properly connect the FPGA within the framework of a computer system. As mentioned in the introduction to this lecture, the FPGAs in reconfigurable computing often work under the direction of a host CPU in a slave-master fashion.

First of all, FPGAs require having access to local memory resources in the form of a local RAM memory. In this context, local memories in the FPGAs are conceptually equivalent to a cache in a traditional ASIC CPU. Second, an FPGA requires ports for input and output (I/O). These I/O ports connect the FPGA to a host CPU. The programming of an FPGA and the numerical inputs required for computations are conveyed through the I/O ports that connect the FPGA to a host CPU and external memories.

Even more sophisticated FPGA architectures are currently available in the market. For instance, FPGAs may include processor blocks and functional components that automatically enable sophisticated arithmetic functions that are commonly used to solve computational problems. In particular, it is not rare to see in modern FPGAs the inclusion of integer multipliers.

1.6 RECONFIGURABLE PROGRAMMING AND FPGA CONFIGURATION

Clearly, reconfigurable programming requires the FPGA to be configured to perform the desired computations. Thus, a *reconfigurable program* is executed by a traditional CPU, which configures and harnesses an FPGA. In this regard, every single logic block and routing path in the interconnect found inside the FPGA can be controlled by using a sequence of memory bits. An *FPGA program* may be viewed as a binary file with a series of memory bits that are used to configure, bit by bit, the logic blocks and interconnects inside the FPGA. The FPGA program is often referred to as the *bitstream*, and it is only a part of a reconfigurable program.

Let us first consider the case of the logic blocks. If we wish to program a single logic block inside an FPGA, then we need to specify all of the programmable points in the architecture of the logic block. By looking at Figure 1.3, we can identify these components and how many programming bits are required:

- We require 16 bits to program the 2^4 possible output values of the 4-LUT.

- The select signal of the multiplexer requires 1 bit.

- The initial state of the D flip-flop requires 1 bit.

Therefore, the programming of a single 4-LUT logic block inside an FPGA requires at least 18 bits. In most practical applications, SRAM (Static Random Access Memory) bits are connected to these programmable points and provide the means to program the 4-LUT logic block. That is, the program for a single 4-LUT logic block is a sequence of 18 bits that uniquely specify all of its programmable points.

Let us now consider the interconnect. In order to program the interconnect of an FPGA, each and every single switch point in the interconnect structure has to be defined. That is, the programming of the interconnect describes how all the logic blocks are routed and connected among themselves. As in the case of the logic blocks, SRAM bits are commonly used to configure the wiring paths of the interconnect.

From a user's perspective, the programming of an FPGA is similar to the programming of a traditional CPU. That is, through a compilation stage an algorithmic description is ultimately transformed into an executable file. However, the process required to program a FPGA is slightly more sophisticated, involving six basic steps:

- Functional Description.

- Logic Synthesis.

- Technology Mapping.

- Placement.

- Routing.

- Bitstream Generation.

A brief description of what is involved in each of these steps is given.

1.6.1   FUNCTIONAL DESCRIPTION

The desired functionality of an FPGA is described using a *Hardware Description Language* (HDL) [7]. The most important examples of these languages are VHDL and Verilog (VHDL is discussed in slightly more detail in a subsequent chapter). HDLs are languages with a syntax and semantic structure that resembles traditional programming languages such as FORTRAN or C. They are used to provide a high level logical description of the algorithm, without invoking specific gates or hardware components. Specifically, HDLs describe the intended behavior of the FPGA under specific inputs.

1.6.2   LOGIC SYNTHESIS

As its name suggests, this step transforms the behavioral description of the FPGA provided by the HDL into a list of interconnected logic gates. The output of this process is often referred to as an *FPGA netlist*. Such a logical synthesis is not unique. For example, we could have fast logical circuits that require a large amount of gates or smaller circuits that run at slower rates. In general, faster circuits require larger area (amount of logic blocks). To this end, most of the synthesis tools available in the market provide the option of optimizing speed vs. area. Clearly, speed and area optimizations play a decisive role in determining the topology of the logical components.

1.6.3   TECHNOLOGY MAPPING

In this step, the individual logic gates described by a FPGA netlist are separated and grouped together into an architecture that best matches the resources of the specific FPGA under consideration [11]. For example, if a simple netlist describes a logical circuit made of two XOR gates and two AND gates, then the technology mapping process determines how to group these four gates in a way that best matches the logical resources available inside the FPGA. As a consequence, the technology mapping process is specific to the type of FPGA that will be used to run the reconfigurable application (on the other hand, logic synthesis was a technology-independent process). Technology mapping tools may be targeted to optimize area, timing, power, or interconnectivity.

1.6.4   PLACEMENT

In this step, all of the groups of logic gates determined by the technology mapping process are assigned to the specific logic blocks available in the target FPGA [3, 28]. Placement is a challenging aspect of FPGA programming due to the exponential number of possible placements: If an FPGA has n logic blocks, then there are n! possible placements. Furthermore, this design has to satisfy some timing constraints, e.g., the time that the FPGA signal propagates from the input to the output registers. As n may well be on the order of millions for modern FPGA architectures, exhaustive search for the most optimal placement is clearly not under consideration. Furthermore, it is not always possible to route any given placement using the available interconnect resources.

1.6.5   ROUTING

Once the placement mapping has been determined, the routing determines the optimal interconnection of the logic blocks using the available resources of the interconnect. Simulated annealing [56] and partitioning [57] are among the algorithms used by vendors such as Altera and Xilinx.

1.6.6   BITSTREAM GENERATION

The final stage generates a binary file that configures all the FPGA logic blocks and interconnects [17]. The resulting bitstream can then be loaded into the FPGA. The bitstream configures the FPGA based on the behavioral specifications that were originally described by the HDL. As such, it is convenient to think of the bitstream as the executable file of the FPGA program and the HDL as the source code.

1.7   BITSTREAM SYNTHESIS

From a user's perspective, given an HDL description, the synthesis of the bitstream is often reduced to a single operation. To date, there are a variety of FPGA vendors that offer sophisticated tools that automatically perform the logic synthesis, technology mapping, placement, routing, and bitstream generation using a single command. Most of these tools are called using a makefile that is based on a directory tree structure.

Unfortunately, even after extensive use of optimizations and heuristic methods, logic optimization, technology mapping, placement, and routing are NP-hard problems. Indeed, the optimization of a superlinearly large number of combinations requires a super-linearly large amount of time. Consequently, bitstream synthesis is a time consuming process that may require a long time to complete. For example, a simple design that performs vector addition of multidimensional vectors may require dozens of minutes for the synthesis of the bitstream to be completed.

1.8   THE XILINX FPGA

Before concluding this chapter, we discuss one of the most popular FPGAs currently available in the market. Xilinx is one of the largest companies that manufacture FPGAs. In addition, Xilinx also develops software and tools to program and harness their FPGAs. Xilinx offers two basic types of FPGAs: The Virtex series for high performance applications and the low performance, but price friendly, Spartan FPGAs. An article to appear in Scientific Computing provides several comparative tables to address costs of using FPGAs [53].

Most reconfigurable supercomputers, like those manufactured by Cray and Silicon Graphics International (SGI), use FPGAs from the Xilinx Virtex family. In addition to the standard FPGA logic fabric, the Xilinx FPGAs feature embedded hardware functionality to carry out some of the most commonly used functions in computational applications. These devices possess a number of integer arithmetic multipliers, adders, and memories. The amount of local memory available on a Xilinx FPGA depends on the specific model and ranges from a few kilobits to tens of megabits.

Xilinx also provides the Xilinx Synthesis Technology (xst) suite of tools that perform synthesis of HDL designs and creates Xilinx-specific bitstreams. The xst tools support both, VHDL and Verilog. Furthermore, Xilinx also provides ModelSim, a rather robust HDL simulation environment that it is extremely useful to verify the functional and timing models of the design before committing resources to the bitstream synthesis.

1.9   SUMMARY

The FPGA is a digital integrated circuit that can be configured and programmed by the user to perform a variety of computational tasks. As we discussed, the basic components of an FPGA are the logic blocks and the interconnects. FPGA programming requires a sophisticated six step process: Functional description of the code, logic synthesis, technology mapping, placement, routing, and bitstream generation. Clearly, reconfigurable programming is much more complicated and time consuming compared to standard programming used for traditional ASIC architectures.

CHAPTER 2

Reconfigurable Supercomputing

Large-scale scientific codes that use reconfigurable hardware and run on supercomputers are often called reconfigurable. In most cases, only portions of the code are implemented on an FPGA while the rest runs on a traditional CPU. The code implemented on an FPGA is usually a computationally intense kernel that is the bottleneck of the application.

In this sense, the FPGA can be considered more as an application accelerator like a GPU (graphics processor unit) than as a replacement for a CPU. A programmer must determine the code segments that are good candidates for FPGA acceleration using such tools as the Reconfigurable Amenability Test (RAT) [54, 55]. This determination should take into account the communication costs between the host CPU and the FPGA application acceleration processor.

Reconfigurable supercomputing combines the strengths of reconfigurable boards with high performance computing [14]. Each node consists of traditional CPUs combined with a high performance FPGA device. A Message Passing Interface (MPI) [44] parallel code can be designed so that an MPI process associated with each node manages its own FPGA. Codes can be run using multiple CPUs and multiple FPGAs.

Reconfigurable supercomputers continue to be developed by mainstream companies including Cray, SGI (now Rackable) and SRC Computers. An SRC-7 cluster system is installed at Jackson State University (JSU) in support of a joint research project between JSU and the U.S. Army Engineer Research and Development Center (ERDC). The High Performance Embedded Reconfigurable Computing (HPERC) market, spurred on by military and space applications, attracts such companies as Nallatech, DRC, and XtremeData. Naturally, various software tools continue to be developed by both hardware vendors such as Altera and Xilinx as well as Intellectual Property (IP) companies such as Mitrionics (Mitrion-C) and Portland Group (PGI) for high-level programming. In particular, PGI is developing compilers to support Compute Unified Device Architecture (CUDA) language to program Graphics Processing Units (GPUs). Reconfigurable codes can be built for each system, but are not portable across platforms from different vendors. The Cray XT5 hybrid supercomputer with FPGA accelerators was introduced in November 2007. We will concentrate our discussion on the architecture and functionality of Naval Research Laboratory's (NRL) Cray XD1 introduced in 2004.

This chapter briefly covers the architecture of the NRL Cray XD1, with particular emphasis on the way the FPGAs are connected to the host processors [20]. How this architecture plays a role in the algorithmic design of the applications that use FPGA acceleration is also discussed. In addition, we will discuss some basic information necessary to understand the FPGA API interface used to establish communication between the host processor and the FPGA.

2.1   NRL CRAY XD1

The NRL Cray XD1 consists of 36 chassis with six nodes in each chassis. Each node of the NRL system consists of two AMD Opteron 275 2.2 GHz dual core processors with 8 GBs of shared memory. Altogether, there are 216 nodes: 144 of them have Xilinx Virtex II Pro FPGAs and six of them have Virtex-4 FPGAs. Even though there are 216 nodes on the Cray XD1, only 150 of them have an FPGA. With this configuration, the NRL Cray XD1 is the largest Cray reconfigurable supercomputer in the world.

The Virtex family of FPGAs are manufactured by Xilinx and, in addition to the standard FPGA logic fabric, feature embedded hardware functionality for commonly used functions such as multipliers and memories. The amount of block RAM depends on the model and ranges from a few kilobits into tens of megabits. Of course, the Virtex-4 FPGA is larger, and its functionality supersedes the one of the Virtex II Pro. The Virtex II Pro (part number xc2vp50-ff1152-7) has a clock frequency of 420 MHz, 23.6 K slices, 4 MB of block RAM and 232 Multiplier Blocks (18x18) [35]. The Virtex-4 LX (part number xc4vlx160-ff1148-10) has a clock frequency of 500 MHz, 67.5 K slices, 5 MB of block RAM and 288 Multiplier Blocks (18x18) [36].

It is important to note that each of the processors inside a node with an FPGA has access to the FPGA inside that node while it has no direct access to an FPGA in a neighboring node. The Rapid Array Interconnect (RAI) uses a bi-directional bus to connect the FPGA with the processors inside the node. The theoretical bandwidth of this connection is 3.2 GB per second (1.6 GB/s each way); however, read operations are much slower at about 10 KB/s due to a flaw in the implementation of the Opteron.

Each of the NRL Cray FPGAs has four QDR (Quad Data Rate) local memories (static RAMs) of 4 MB, which conceptually are equivalent to the cache memory on a standard processor. Each of these QDRs can be read and written simultaneously, allowing a total of eight concurrent memory operations.

The memory bandwidth between the FPGA and its cache is 3.2 GB per second. This large bandwidth means that the FPGA could, in principle, be kept busy while the host processor is concurrently transferring data to and from the FPGA. While the host processor is filling up one buffer in a QDR SRAM, the FPGA can work with data stored in another buffer. Unfortunately, coordination between the host processor and co-processor is a challenge, and some of the tools do not even support data transfers to/from the FPGA while the device is running.

Because the diverse ways to send data to and from the FPGA are asymmetrical with respect to their speed, care needs to be taken when performing I/O operations. Table 2.1 summarizes the I/O channels and their speeds.

Therefore, an efficient reconfigurable application must be designed in such a way that the slow I/O paths are not taken. The design must never involve the FPGA reading data directly from the host RAM. Instead, the host processor writes any needed data into the FPGA QDR memory. Similarly, the host processor should never read results from the FPGA QDRs, and the FPGA should write any needed results directly to the host RAM.

Table 2.1: Approximate Bandwidth of I/O Operations

Source	Destination	Read	Write
FPGA	Local QDR Memory	4x800 MB/s = 3.2 GB/s	4x800 MB/s = 3.2 GB/s
FPGA	Host RAM	N/A (*)	711 MB/s
Host	FPGA QDR Memory	10 KB/s (*)	3.2 GB/s
Host	Host RAM	Fast (GB/s)	Fast (GB/s)
(*) = Slow or not available.			

2.2   CRAY API

After an FPGA program is synthesized, the resulting binary file encapsulates the hardware configuration of the FPGA (see Section 1.7 on bitstream synthesis for further information about this process). In order to load this file into the FPGA, harness the FPGA program and establish communication links between the host processor and the FPGA, Cray offers a utility program and an application programming interface (API) library of functions. Other FPGA programming tools, such as Mitrion-C and Celoxica, offer their own API libraries, which in turn are wrap-ups of the Cray API. Without loss of generality, we will concentrate on the Cray API.

The Cray API is not used to program the logic of the FPGA but merely to harness its functionality and interact with the host. The FPGA code knows nothing about the API. In this regard, an FPGA works like any other device connected to the system. That is, the API is used to open the FPGA and get a handle to the device, load the FPGA program, allocate and initialize memory, to execute the FPGA logic, send and retrieve information from the FPGA, and halt the execution and close the FPGA. All of the API calls are made within a program running on the host processor.

The Cray API provides various calls in C to transfer data to and from the FPGA, map memory, load and unload the FPGA, and open and close the FPGA device [12]. Table 2.2 lists available Cray API calls.

User logic in the FPGA has access to the Opteron memory through the RapidArray Transport (RT) core[1]. The Cray RT core is written in VHDL. This core provides data transfers from the FPGA to the Opteron over the RT interconnect. The user logic in the FPGA sends a bus transaction to the RT core, which forwards it through the RT fabric to hardware on the Opteron where it becomes a read or write transaction to the Opteron DRAM.

Figure 2.1 shows the physical components along with their address spaces [12]. The FPGA memory is accessible via a region of the HyperTransport I/O address space. Specifically, the FPGA memory occupies a 128 MB address region in the Rapid Array Processor (RAP). This memory is addressable from the application running on the Opteron. In the figure, the Opteron's I/O space contains memory maps for RAP-1 and RAP-2, the latter being the interconnect to the FPGA.

---

[1]A core is a "soft" processor embedded in the FPGA logic.

Table 2.2: List of Cray API Calls for Interfacing with an FPGA	
Cray C API Call	**Description**
`fpga_open`	Opens an FPGA device
`fpga_load`	Loads a converted binary logic file into an FPGA
`fpga_reset`	Places the FPGA user logic into reset
`fpga_start`	Releases the FPGA user logic from reset
`fgpa_memap`	Maps a region of the FPGA address space into the application address space
`pga_mem_sync`	Forces completion of all outstanding transactions mapped to FPGA memory
`fpga_register_ftrmem`	Registers a region of application memory for direct access by FPGA
`fpga_dereg_ftrmem`	Deregisters an FPA transfer region
`fpga_rd_appif_val`	Reads a value from the FPGA address space and guarantees access order
`fpga_wrt_appif_val`	Writes a value to the FPGA address space and guarantees access order
`fpga_status`	Gets the status of an FPGA device
`fpga_unload`	Clears the programming of an FPGA
`fpga_close`	Closes an FPGA device

RAP-2 has a 128 MB region that the FPGA RT-core memory-maps to the FPGA. Data from the host can be written to the QDRs, and block and distributed RAM on the FPGA.

The `fpga_memmap`, `fpga_rd_appif_val` and `fpga_wrt_appif_val` calls allow the host application to read and write data to the memory of the FPGA. In particular, an `fpga_memmap` call sets up an address space, which can be accessed by pointers. The `fpga_rd_appif_val` and `fpga_wrt_appif_val` calls read and write, respectively, single 64-bit values at a time. Certain conventions are built into the memory map of the FPGA. For example, the QDR SRAM address space begins at 64 MB, with each QDR address space occupying a contiguous 4 MB region. The `fpga_register_ftrmem` call sets up a host address space to which the FPGA can read and write. The minimum size is one memory page and the maximum size is 1 GB. Section 5.7 gives a detailed example of the Cray API.

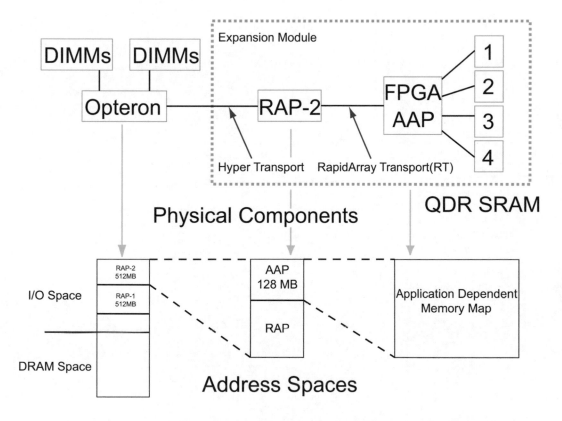

Figure 2.1: Physical components and Address Spaces of the Cray XD1 Opteron and FPGA.

CHAPTER 3

Algorithmic Considerations

By combining software and hardware in a sophisticated architecture, FPGAs offer the prospect of increased performance for a variety of computational applications. Currently, a variety of applications in the areas of cryptography, bioinformatics, signal analysis, and image processing have benefitted enormously from the use of an FPGA as a computation acceleration device [34].

Unfortunately, there is no compiler flag that automatically enables FPGA acceleration. A programmer has the entire burden of designing and rewriting code to make it reconfigurable. In addition, there is no way to predict, at the design stage, if a reconfigurable code will exhibit a dramatic increase in performance. As a matter of fact, it may well happen that the reconfigurable code is slower than the original ASIC application. Indeed, not every computational code will be a good candidate for FPGA acceleration.

Therefore, a software developer must follow a series of algorithmic considerations, timing analyses, and benchmarking experiments to determine whether or not an implementation is a good candidate for FPGA acceleration [14, 15, 53, 57]. Furthermore, because reconfigurable supercomputing is a time consuming endeavor, it is crucial to determine if a specific code is worth the effort at an early stage.

3.1 DISADVANTAGES

Even though it has been shown that FPGAs offer huge performance advantages for a number of computational applications, this technology also imposes a series of unique restrictions and limitations. This should not come as a surprise. FPGAs are not nearly as sophisticated as ASIC CPUs. As a consequence, a programmer has to determine the best way to leverage the advantages and disadvantages offered by reconfigurable supercomputing.

The main disadvantages of FPGAs regard their speed, data transfer overheads, data types, and level of effort required, which we discuss next.

Speed

Currently available, FPGAs are usually much slower than most ASIC boards. For example, while most of the modern CPUs work in the range of 3 GHz, connected FPGAs run on the order of 350 MHz. Thus, FPGAs are an order of magnitude slower than ASIC CPUs. Nevertheless, FPGAs are becoming faster due to fabrication improvements. Importantly, having a slower clock allows FPGAs to run using approximately $1/10^{th}$ of the power and cooling requirements of their ASIC counterparts [58].

Data Transfer Overheads

When programming reconfigurable codes, the memories of the processors and coprocessors are rarely shared. Hence, it is normally necessary to transfer data from the memory of the CPU to the local memory of the FPGA, which delays computations because the coprocessor must wait for the transfer to complete. Similarly, the FPGA device needs to write the results to the memory of the host processor, which can be done concurrently while the coprocessor is computing other results. Depending on the specific architecture of the reconfigurable computer, this movement of data to and from the FPGA may take a significant amount of time. Loading the logic onto the device takes by far the most amount of time. As a consequence, this overhead could override any improvement in the computational performance provided by the FPGA.

Data Types

Even though floating point operations are becoming more feasible to implement in modern FPGAs, the most efficient implementation, implementations use integers or fixed point numbers [32]. Indeed, the use of a small number of floating point variables may be enough to use most of the logic blocks available in the FPGA. Increasing the bit width adds logic over most of the area of the chip as numerous LUTs are needed.

Level of Effort

Reconfigurable computing using FPGAs requires a large investment in time for design, programming, simulation, testing and debugging. There is a learning curve for any tool. The tools, which are themselves new technologies, must adapt to the rapid changes in FPGAs and systems which use them. Because reconfigurable supercomputers are complex, many different software tools are needed. Although these divergent tools largely work well together, none of them alleviate the programmer of much of the hard work, especially in the area of design.

3.2 DATA PARALLELISM, PIPELINING, AND CONCURRENCY

As mentioned in the previous section, FPGAs suffer from a series of disadvantages. However, it is important to understand that the real advantage of FPGAs is that, depending on the application, they can be configured to exploit massive amounts of parallelism. In this regard, ignoring the time needed to load the logic and the overhead of data transfers, if an FPGA is ten times slower than a CPU, then the FPGA needs to perform at least ten times more work per work cycle than the CPU, so both are of comparable performance. To be worth the effort, it is desirable that the FPGA will perform at least 100 times more work per cycle. In such a case, the transition of an implementation to a reconfigurable architecture is accelerated by a factor of ten, which justifies the time and resources spent in software development.

However, an increase in performance by a factor of 100 is only possible if the original application features a high degree of data parallelism, pipelining, and concurrency.

Data Parallelism

As in traditional supercomputing, data parallelism [47] refers to the distribution of data across different computational nodes. In this regard, FPGAs work best with datasets that only exhibit a few, if any, data dependencies. Indeed, FPGAs are good at exploiting data parallelism because the same operations can be performed on different data. As a consequence, the FPGA design will require a significantly smaller amount of logic blocks and interconnect resources. With data parallelism, the same logic blocks can be used over and over again until all the data has been processed. Also, the operations can be reordered to improve computational performance. Furthermore, computations exploiting data parallelism can be scheduled to maximize data reuse, to increase computational performance, and to minimize memory bandwidth.

Pipelining

As in traditional supercomputing, pipelining is related to being able to overlap computations. In this case, overlap usually involves shifting operations in time so that different pieces of hardware are working on different stages of the same task. Pipelining leads to an effective level of parallelism in the implementation. Data parallelism cannot always be fully exploited due to I/O limitations; whereas, pipelining permits parallel processing without increasing I/O bandwidth.

In essence, pipelining is particularly important in reconfigurable computing because the operations carried out in the FPGAs are often limited by delays in the interconnect. Therefore, pipelining is essential to increase parallelism and reusability of the reconfigured hardware at the expense of some latency (the amount of time it takes for a block of data to be processed). As a consequence, target applications that are able to tolerate latency are suitable candidates for FPGA acceleration.

Concurrency

As in traditional supercomputing, concurrency is the ability of certain codes to perform several different computations simultaneously by different computational units. Thus, if we have enough data parallelism in the dataset, then we can apply a large amount of logical operations at the same time. In other words, we exploit task parallelism. Even though FPGAs are substantially slower than ASIC CPUs, their advantage resides in the amount of instruction level parallelism that they offer. Then, in order to obtain increased computational performance, the FPGA needs to complete at least 50 to 100 times more instructions than a regular CPU.

3.3   ALGORITHMIC REQUIREMENTS

In addition to high levels of data parallelism, pipelining, and concurrency, the target application must conform to a series of algorithmic considerations for maximum acceleration. These considerations involve data element size, arithmetic complexity, and control structures.

Data Element Size

The size of the data element that is processed by the FPGA determines to a great degree the speed and size of the reconfigured circuit. This statement is true regardless of the specific data type. Indeed, if the application offers a high degree of data parallelism, then the computational performance will depend on how many operations can be performed concurrently. Larger data size elements lead to larger circuits, allowing fewer computational FPGA elements and thus less parallelism. Therefore, one of the goals of high performance reconfigurable supercomputing is to look for an efficient data representation that uses the fewest possible number of bits.

Arithmetic Complexity

Similarly, as with large data elements, complex arithmetic operations require larger circuits, reducing the number of processing elements available and impacting the degree of parallelism. In order to obtain the highest performance, it is desirable to use the simplest operations possible. Consequently, reconfigurable computing will often imply a precision/performance trade-off.

Control Structures

In general, FPGAs have much better computational performance if most of the logic operations can be statically scheduled. Indeed, from the context of a computer program, it takes time and resources to make decisions. Furthermore, the conditional **if** statement requires sophisticated circuitry that has to be implemented by the logic blocks and interconnect in the FPGA. Thus, static control structures may significantly speed up the computation time and require a considerably smaller amount of logic resources. In this regard, datasets that exhibit few dependencies often require simple control requirements.

3.4   I/O CONSIDERATIONS

As previously argued, the power of reconfigurable supercomputing using FPGAs resides on the exploitation of data parallelism, pipelining, and concurrency. However, in order to fully exploit concurrency and pipelining, data must be transmitted at high enough rates to keep the hardware busy.

If there are more I/O than compute operations, concurrency and pipelining will only have a small effect on the overall performance of the reconfigurable application. In such a case, increase performance will require higher memory bandwidth, which may or may not be available. On the other hand, if there are more compute than I/O operations, then concurrency and pipelining will be able to accelerate the application by a considerable amount.

Optimal FPGA acceleration takes place by carefully scheduling data transfers and computation in order to achieve substantial levels of concurrency and pipelining. Data elements that have been fetched from memory should be reused multiple times. In other words, it is desirable to keep many processing units busy while using the exact same data.

As a consequence, limited I/O bandwidth in the FPGA device imposes serious restrictions on the type of algorithms that can be accelerated using reconfigurable computers. If an algorithm involves considerable I/O (it is an I/O bound algorithm), then it may not be a good candidate for FPGA acceleration.

On the other hand, computation bound algorithms are very good candidates for FPGA acceleration. Of course, the type of operations being performed limits the degree of improved performance. Regardless, the potential speedup depends on the exploitation of data parallelism, pipelining, and concurrency.

3.5   LOOP STRUCTURE

Any Boolean function can be mapped to an FPGA. The computation of such a Boolean function will be more efficient in the FPGA if we are able to exploit data parallelism, pipelining, and concurrency. If an algorithm invokes a computational kernel that repeats itself multiple times inside a loop, then the kernel could be programmed inside the FPGA. Therefore, the ASIC CPU executes the program and only invokes the FPGA to execute the compute-intensive loop.

Programs that are good candidates for FPGA acceleration include those for which a single loop is performing a large number of simple operations. In these codes, the loop dominates the overall execution time of the algorithm. In this case, the loop and its kernel are the targets for FPGA acceleration, instead of the entire code. Furthermore, by Amdahl's law, the acceleration of this single loop will lead to a significant increase in the overall performance of the entire code.

Similarly, if the target code is rather complex, featuring dozens of time consuming loops and other bottlenecks, then the FPGA acceleration of a few loops will not provide any significant gains in performance. Indeed, from Amdahls' law, the acceleration of a few loops will not provide any significant gains in computational performance.

Even if a single loop dominates the execution time of the target program, further algorithmic considerations are necessary before considering the use of FPGAs to accelerate the code. In general, the structure and type of operations inside the loop are dramatically restricted. Issues involved in the analysis of the loop include:

- Avoid pointer operations used to handle common data structures such as linked lists. Such data structures are inconsistent with parallel access and efficient use of local memory. In general, an implementation must be rewritten, if necessary, to handle only values instead of pointers.

- Avoid double precision computations, as well as sophisticated trigonometric, logarithmic, and transcendental operations. The reason is that these operations require a considerable area of the chip.

- Avoid complex data dependencies. Recall the goal of FPGA acceleration is to take advantage of the potential parallelism in the implementation. Hence, the body of the loop must be easily parallelizable. Complex data dependencies typically require some of the configured blocks of hardware to remain idle until data becomes available.

- The number of iterations should be known at compile time. Indeed, runtime control requires sophisticated logic structures that demand large amounts of logic blocks and interconnects.

- Data blocks of arrays accessed in the body of a loop should not be very big. FPGAs execute more efficiently by applying operations to a relatively small number of data elements. In such a manner, an FPGA is able to reuse large portions of logic blocks and interconnects. It may be possible to rearrange the computations to permit such data accesses. Important considerations include:

  - I/O requirements. For instance, searching a large database using FPGAs is extremely inefficient unless the host processor and coprocessor enjoy shared memory.

  - Depth of nested loops. The implementation of four or more nested loops will consume a large amount of logic blocks and interconnect. Also, deeply nested loops will negatively affect latency.

  - Number of conditional operations. The exact same operations should be performed on all of the elements inside a loop. Conditional statements inside a loop are perfectly correct and allowed by the model. However, conditionals will consume valuable resources and a large number of conditional statements may become a limiting factor for successful acceleration.

A clear pattern emerges from these loop restrictions: In order to achieve considerable computational gains, the loops should be easily expressed in an algorithm that exploits data parallelism, pipelining and concurrency.

3.6   EFFECTIVE RECONFIGURABLE ALGORITHM DESIGN

At this point, it should be clear that FPGAs and CPUs offer distinct capabilities, restrictions, advantages, and disadvantages. Therefore, an algorithm that may be a poor choice for a traditional CPU may actually be an excellent choice to be implemented on a FPGA, and vice versa.

For example, consider an application that merely organizes data into a histogram. That is, given a dataset, we gather data elements into specific groups according to their value. In traditional C programming, for instance, an optimal implementation does not involve the brute force approach of exhaustive search. Instead, it is recommended to implement a variant of binary search. On the other hand, an FPGA implementation based on the brute force method is practical as its inherent data parallelism, concurrency, and pipelining attributes can be directly exploited.

Let us consider another example. Suppose the following code segment is the kernel of a program that we wish to accelerate using FPGAs.

```
for (i=0; i<n; i++)
{
    a[i] = 0;
    for (j=0; j<m; j++)
        a[i] = a[i] + K[i][j]*f(a[i]);
}
```

Is this computational kernel a good candidate for FPGA acceleration? Following our previous discussion, we can conclude that, if n and m are known at compile time, a and K are arrays of integers or single precision numbers, and f is a function that does not involve trigonometric or logarithmic operations, then this loop appears to be a good candidate for FPGA acceleration. Various strategies are possible to implement this nested loop in an FPGA; some of which we discuss next.

Sequential

All the computations are performed inside a single processing element. This strategy does not exploit any possible level of data parallelism, instruction concurrency, and pipelining. As FPGAs are considerably slower than CPUs, and due to the big data transfer overhead involved, this strategy will perform much worse than the original CPU implementation.

Parallel

This strategy performs the loop computations using many processing elements. That is, the reconfigurable circuit that computes the body of the loop is replicated several times across the entire area of the FPGA. Using a completely parallel strategy, all $O(n \times m)$ loop operations are performed simultaneously, in a single computational step. This strategy clearly exploits the available degree of data parallelism and instruction concurrency. However, it may require a large amount of logic blocks and interconnect resources, which may not be available.

Partly Parallel

This strategy uses $O(n)$ processing elements to compute the body of the innermost loop. Therefore, the entire code is completed in about $O(m)$ computational time steps. This strategy is the best of the three choices because it reuses hardware with good performance.

3.7   SUMMARY

To summarize this section, we provide a simple heuristic methodology to determine if a given code is a good candidate for the FPGA:

1. Perform benchmarking analysis to determine if the code is computation or I/O bound. If it is I/O bound, then it is probably not a good candidate for FPGA acceleration.

2. Perform benchmarking analysis to determine any possible bottlenecks in the code. If most of the time is spent in a single loop, then this loop is a good candidate for the FPGA. If there are many loops consuming most of the computational time, by Amdahl's law, it is unlikely that we will observe significant improvements using an FPGA implementation.

3. Determine if the target loop exhibits sufficient opportunities to exploit data parallelism, concurrency, and pipelining.

4. Perform algorithmic analysis to address all the algorithmic considerations described in this chapter, such as parallelism, loop structure, I/O, etc.

CHAPTER 4

FPGA Programming Languages

One of the most challenging aspects of reconfigurable computing is the actual software development for a particular FPGA variant. FPGAs natively require Hardware Description Languages (HDL) such as VHDL or Verilog. VHDL and Verilog, were specifically developed for hardware design. As a consequence, these languages require enormous effort to develop software applications based on either existing implementations or algorithms. Thus, VHDL and Verilog are exceptionally good languages for the description and design of logical circuits, but they are extremely inefficient for reconfigurable computing of scientific problems.

Such deficiency is made evident by surveying the actual literature on VHDL and Verilog programming. Most of the published books on the topic consider examples geared to implementing digital devices such as multiplexers and decoders. But there is no substantial discussion or examples illustrating how to use these languages to perform the programming of a search or sorting algorithm. As such, it becomes a challenge for the software engineer to develop computational applications using VHDL or Verilog.

In recent years, a handful of FPGA programming tools and languages have appeared in the marketplace. These products offer to ease FPGA programming effort by using similar syntax and semantics as with traditional programming languages like C and Java. A few others employ user-friendly interfaces of well-known computational products such as MatLab. The compilation of these programs produces a VHDL file that can be used to synthesize a bitstream using the FPGA vendor's bitstream synthesis tools.

In this category, some of the most popular FPGA programming tools and languages include DSPLogic [48], Handel-C [25], Impulse-C [49], and Mitrion-C [21]. Each of these languages presents its own specific set of advantages and disadvantages. In general, the common advantage these products offer is that their syntax and semantic structure is better suited for algorithmic design than for hardware description. Handel-C, for instance, has the advantage that some of its semantics and syntax are identical to traditional C. On the other hand, DSPLogic is based on Matlab and Simulink, and uses a very simple graphical user interface. Matlab and other packages such as Mathematica, Star-P, and Python are considered Very High Level Languages for HPC [41].

Nevertheless, these FPGA tools present a variety of obstacles. Each tool requires a programmer to learn a new paradigm. Although an experienced programmer can learn a traditional language in a couple weeks, it takes much longer to understand a new paradigm as well as the syntax and semantics. In particular, good parallel programming is challenging by itself.

Furthermore, reconfigurable acceleration of a code with a limited tool kit is a formidable task even for an advanced FPGA programmer. In addition, because these languages basically describe the functionality of the FPGA logic blocks and interconnects, they are severely restricted in the types of things that they can do. These tools are less efficient and less flexible when compared to a low level HDL design. Notwithstanding, we expect compilers will increasingly map higher-order codes more efficiently onto FPGA devices.

Regardless of the languages or tools used, the programming of FPGAs is a challenging and time-consuming task. In this regard, the principal problems confronted by the software developer include a difficult debugging process and a nearly total lack of portability. Efficient debugging, for instance, is extremely difficult to accomplish within the realm of reconfigurable computing. The principal reason for this limitation is the inability to easily print the contents of variables in the FPGA code without increasing the logic and changing the timings, i.e., without changing the program. In addition, the bitstream synthesis is a process that typically takes several hours or more. Hence, the traditional debugging strategy of using multiple print statements and frequent recompiling is not feasible when programming FPGAs.

Within the context of reconfigurable computing, the only way to perform debugging is using a simulator provided by the vendor. These simulators mimic the functionality of the FPGA under certain inputs and provide a reasonable way to determine when an error has occurred. The usefulness of an FPGA programming language is limited by functionality that the associated simulator provides.

Further complicating the FPGA programming process is the lack of code portability across different reconfigurable supercomputers. For example, the SGI and Cray reconfigurable supercomputers require completely different host and FPGA code to account for their differences in terms of architecture and available FPGA devices.

Furthermore, there is no portability across different FPGA programming languages. For instance, Mitrion-C code is completely different from DSPLogic, Handel-C, Impulse-C, and VHDL. In addition, the specific interface to the host CPU is also completely different for all of these FPGA programming languages.

In the remainder of this chapter, we will closely examine four popular FPGA programming languages: VHDL, DSPLogic, Mitrion-C, and Handel-C. Other languages such as Verilog, Impulse-C, and Chimps are covered elsewhere.

4.1 VHDL

Register Transfer Level (RTL) tools provide a description of a digital design using logical expressions. Thus, RTL describes behavior of a digital circuit in terms of signals between registers and the logical operations performed on them. In this regard logical expressions, such as AND or XOR, form a higher level of abstraction above gate description. RTL tools allow the full control of register-to-register logic without having to select the specific gates that implement the design.

As a consequence, RTL tools are helpful to create a high level representation of a circuit for which a low level gate and circuit representation can be easily derived. In particular, Hardware Description Languages (HDLs) produce an RTL abstraction of digital circuits. HDLs focus on the description of how signals flow between different registers and how they change under certain logic operations. Examples of HDLs include VHDL and Verilog.

VHDL stands for VHSIC Hardware Description Language where VHSIC is Very High Speed Integrated Circuits. VHSIC was a project sponsored by the US Department of Defense during the 1980s to accelerate advances in digital circuit technology. VHDL, as an RTL hardware description language, is a product of the VHSIC effort. Thus, VHDL was conceived as a tool to document the behavior of ASIC processors. VHDL offers a superior alternative to layout, focuses on the functionality of the device, and is completely independent of details specific to the implementation [23, 24].

In particular, one can design digital hardware in VHDL [7]. The description of the design in VHDL is subsequently used to produce the RTL schematic of the digital circuit. In addition, VHDL allows the digital design to be modeled and verified before the synthesis tools translate the design into a specific implementation using wires and gates.

Some of the features that distinguish VHDL include the following:

- Strongly typed.

- Based on Ada.

- Features several similarities with object oriented languages such as C++ and Java.

- Contains special constructs and semantics to describe instruction concurrency at the hardware level.

- Follows IEEE standards.

- Periodically revised to reflect the latest technological and industry trends.

Some of the deficiencies of VHDL as a programming language for FPGAs, within the context of reconfigurable supercomputing include the following:

- The user is required to learn and master VHDL, a time consuming effort.

- VHDL is algorithmically deficient.

- VHDL requires several lines of code to describe very simple arithmetic expressions.

- The functionality of the FPGA is described at a very low level of abstraction, which may not be of interest to the developer of reconfigurable computing codes.

- VHDL as a language to program FPGAs inside reconfigurable supercomputers lacks portability. Indeed, the VHDL code strongly depends on the technological implementation of the specific FPGA.

- VHDL is extremely difficult and time consuming to benchmark and optimize.

• VHDL code is extremely difficult to debug.

VHDL is a language that was developed especially for digital circuit design and not for algorithmic programming. This observation makes it clear why VHDL is mainly used by engineers to handcraft special implementations and not by programmers accelerating codes within the realm of reconfigurable supercomputing. As a consequence, unless the user is an expert VHDL programmer, this language is extremely difficult for the development of algorithmic applications. Consequently, unless the user is already an expert VHDL programming, it is not recommended as a language to program FPGAs.

Verilog has similar capabilities as VHDL. Perhaps the most notable difference is that Verilog is not as strongly typed as VHDL. And truth be told, supporters of both languages regularly sustain heated debates concerning the superiority of one over the other. In any event, for the purposes of reconfigurable supercomputing, Verilog and VHDL are equivalent in the sense that they both share the same type of shortcomings for software development.

4.2   DSPLOGIC

DSPLogic produces a "Rapid Reconfigurable Computing Toolbox" aimed at the graphical programming for the fast development of computational applications on FPGAs and standard CPUs. In strong contrast with Mitrion-C and Handel-C, DSPLogic does not offer a new computational language with syntactic and semantic similarities to C, but presents, instead, a sophisticated graphical interface where a computational program is built by interconnecting a variety of modules.

The DSPLogic system is based on Matlab and Simulink, two popular software packages for mathematical analysis and simulation. Matlab provides a powerful numerical computing environment harnessed by an interpreted, high-level programming and scripting language. The three obvious strengths of Matlab are its implementation of linear algebra, Fourier analysis, and numerical algorithms. Thus, Matlab is widely used for a variety of applications in the areas of optics, radar, acoustics, and signal analysis.

Simulink is a graphical language that wraps Matlab into a sophisticated graphical block diagramming tool. A large number of computational modules are included in the block library of Simulink, and permits the user build, customize, and store new modules. In Simulink, algorithms are implemented by dragging library blocks into a visual programming workspace and establishing connections between them. Simulink also adds a time flow variable, which permits the simulation of the temporal evolution of the system. By relying on the computational features of Matlab, Simulink is a rather powerful tool for modeling, simulation, and analysis of multidomain dynamical systems. Simulink is mostly used in control theory and digital signal analysis.

DSPLogic's RC Toolbox is fully integrated with Matlab and Simulink, inheriting both of its advantages and disadvantages. Thus, FPGA programming using this tool must be done using Simulink's graphical environment, with the benefit that all FPGA implementation operations are hidden in layers of abstraction inside library modules. The library includes modules that in principle allow access to Xilinx primitives and VHDL code. The program can be designed to run sequentially, pipelined, or in parallel.

Because of its tight integration with Simulink, DSPLogic RC offers very efficient and reliable simulation and debugging tools, however, DSPLogic works only on a Windows-based PC. The user has to design, program, and synthesize the FPGA code on the PC and then move the resulting bitstream file to the XD1.

The data streaming and communications between the host processor and the FPGA are handled by the DSPLogic's messaging abstraction layer, a wrap of the Cray XD1 API library functions. The host program invokes a series of library calls to establish the connection, configure the device and communicate with the FPGA.

Overall, DSPLogic's RC diagrammatic programming appears to be better suited for digital circuit design although it has been used for bioinformatics, cryptography, and signal processing. The block abstraction and code encapsulation features inherited from Simulink may prove valuable for the design of very large and complex reconfigurable codes. However, even simple reconfigurable codes, such as the addition of two multidimensional vectors, often require dozens of interconnecting blocks. In addition, based on the Simulink graphical programming paradigm, the user needs to learn Matlab and Simulink.

4.3 HANDEL-C

Handel-C is a unique Hardware Design Language (HDL), which was originally developed by Oxford Hardware Compilation Group and was later developed into a proprietary product by Celoxica (acquired by Catalytics, which was merged to form Agility whose C synthesis assets were eventually acquired by Mentor Graphics in 2009), which is no longer supported. In any case, Handel-C is well established compared to the relatively new and novel language Mitrion-C (covered in the next section). Handel-C may be viewed as an extension of a large subset of standard C, including its precedence rules for operators and syntax for control structures (`for`, `if`, `switch`, `while`), functions and expressions, operators, and data structures (arrays, structures, pointers, variables) [25]. The name honors the classical music composer George Frederic Handel, with the "-C" indicating its relationship to the C language.

Handel-C code is intended to be portable excluding the parts dealing with low-level hardware interfaces (pins) and customized interfaces to software modules written in Handel-C, VHDL or Verilog (ports). Yet, simply redefining the interfaces for another system is not likely enough to obtain a program that runs satisfactorily. Indeed, writing an efficient program normally begins with a careful design of the interfaces to effectively utilize the hardware. On the other hand, functions compiled into libraries can be reused on the same system or on the same platform with identical I/O ports and memory subsystems.

The language provides channels for point-to-point communication between processes executing in parallel. Channels are familiar from the occam programming language that is based on Communicating Sequential Processes (CSP). One branch outputs data onto a channel and another branch reads data from the same channel. If there are multiple channels, reads and writes are listed according to a desired priority to select the first channel that is ready (prialt). If communication is not synchronized, then a FIFO queue is employed so that writes are blocked only when the queue is full, and reads are blocked only when the queue is empty.

The language also provides semaphores to allow the coordination of shared resources. A semaphore should be held only as long as needed (trysema and releasesema) in the "critical section" such as an update of a variable that is shared. Priority is undefined (as its implementation is proprietary) when several processes concurrently attempt to take a semaphore to create a lock.

Each assignment, increment, decrement, channel communication, delay, and release of a semaphore takes one (clock) cycle. Functions, expressions, conditions and jumps (`break`, `goto`, `continue`, `return`) are evaluated in zero cycles. Signals are objects that behave like wires as during the same cycle the value read is the value assigned.

Loops require at least one cycle; whence, empty loops are not permitted. A conditional test (`if` block) before a loop is inserted whenever there is a possibility the body is not executed at least once. Whenever inserting an **if** block, an **else** block is included with a delay statement. Similarly, a delay is added in a default statement of a switch construct. The compiler may also add delay statements to break loops in order to avoid timing conflicts.

For loop's assignment of the loop counter after the execution of the loop body requires an extra cycle. While loops are preferable as the loop counter is updated in parallel with other statements in the loop body. This optimization is significant whenever the entire loop is executed many times (such as in a nested loop).

At most one "element" in an internal block or external ROM or RAM can be accessed in any clock cycle. ROMs permit only read accesses. By using "multi-ported" RAM (mpram), it is possible to simultaneously perform a read and a write, a couple reads, or a couple writes. The maximum number of accesses in mpram during a cycle depends on the actual number and type of ports available in hardware, e.g., a Virtex-4 FPGA has a read/write port and a read port.

The clock period is determined by the longest expression (involving arithmetic and shifting operations and possibly signals) in a single statement, the deepest depth of nested control structures, or the longest chain of logical control blocks. The reason is that all of the operations or conditions must be executed in a single cycle, which means the delay can become significant. If the delay is too long, timing constraints may not be met. A programmer should avoid division, break up complex statements into several simpler statements (using registers instead of signals), reduce the bit-widths of variables, and avoid nesting control structures more than a few layers deep. If the code is carefully written using registers in a pipeline fashion, flip-flops can be automatically moved by the compiler.

Execution of successive statements and blocks is sequential by default, except all cases of a switch statement (and all branches of an if block) are evaluated in parallel. Handel-C provides a par(allel) block to concurrently execute all statements and blocks within the par block. Care must be taken when grouping statements together in a par block because it is not possible to write to the same variable during the same cycle. Blocks may be explicitly modified as seq(uential) blocks (default case) as documented. Parallel and sequential blocks may be nested. Parallel blocks complete only when all statements and blocks within the par block complete, i.e., when the longest "branch" completes.

Handle-C also provides parallel (`par`) and sequential (`seq`) loops. The loop body is duplicated with appropriate indices depending upon the number of steps specified. In par and seq loops, the replicated bodies are executed in parallel and sequentially, respectively.

Side effects are not permitted in expressions. In particular, the comma operator cannot be used to list expressions. Consider the following statement in C:

```
a = d--, --b - 1;
```

This C statement may be written in Handel-C as

```
par
{
    a = b - 2;
    b--;
    d--;
}
```

Variables are in effect registers. It is not safe to assume a variable is initialized to zero unless it is static. Only static and global variables can be initialized. Initialization reduces the amount of logic and clock cycles needed.

The width of data types is not fixed (not restricted to 16-bit, 32-bit, or 64-bit, etc.). The width of a variable is either specified or automatically inferred. Avoid using longer widths than necessary to more efficiently utilize the hardware resources on the chip.

Unlike in standard C, a Handel-C program may consist of multiple `main()` functions, each of which is associated with a clock so that different parts of the program run at different speeds. However, all `main()` functions in the same source file must use the same clock. Unlike other functions, a `main()` function neither takes any parameters nor returns any values.

At most a single function call may appear in and only in an expression statement. For example, `f(x)` is a valid expression; whereas, `f(g(x))` and `f(x)+g(x)` are invalid expressions. Function parameters (including pointers) are copied from the calling function, i.e., references are not modified. The types of the parameters must be explicitly declared in the function definition.

Functions may be either inlined for parallel execution so that the hardware resources are not shared, or "shared" for sequential execution. Arrays of functions may be defined to make a fixed number of copies for parallel execution. Whenever a function is copied (whether in an inline function or an array of functions), all statically declared variables are independent among all of the copies. Recursive functions are not supported because all of the logic must be determined at compile time.

The language provides extensive macro support to avoid rewriting tedious details. Each macro call involves unshared hardware resources. Macro expressions are useful to make copies of expressions (vs. statements) for parallel execution. Unlike macro expressions, a macro procedure, which is similar to an inline function, consists of complete statements and does not return a value.

Unlike functions, macros permit call-by-reference untyped parameters so that the actual references are used instead of a copy. For instance, a macro procedure swap(x,y) swaps the values of the actual variables x and y listed in a macro call whereas a function call swap(x,y) cannot modify the references (only copies of x and y). Non-parameterized macros are similar to preprocessor directives (#define).

Complex macros may be defined using "let" definitions of simpler macros. Recursive macro expressions and procedures are supported. A "select macro" works like the ternary operator (?:) in C except the condition is evaluated at compile time; whence, the hardware is not configured to perform the ternary operation. Similarly, an "ifselect macro" works like a logical if statement: Only the selected block of code is expanded at compile time.

Unlike macro expressions, shared expressions provide opportunities to reuse hardware to reduce the amount of time that resources are idle. Shared expressions are useful when the same computations (involving different inputs) are performed at different times during program execution. Shared expressions are defined and called using the same syntax as macro expressions except for the declaration as a shared or macro expression. Although recursion is not supported directly, recursive shared expressions may be built up using recursive macro expressions.

The language provides built-in base data types for characters, signed and unsigned integers. Derived types for data aggregates include arrays and structures. Arrays are like RAMs as a linear memory model. Arrays are needed for parallel access due to the restrictions for accesses to memory. Structures are useful for holding data like the mantissa and exponent of floating point numbers. Support for fixed point and floating point is provided in the form of libraries. In addition to familiar operators, the language provides special operators for bit manipulation, range selection (a[MSB:LSB]) and concatenation.

Handel-C provides buses for interfaces with external devises using pins. User-defined interfaces are used for communication involving other codes when Handel-C is the top-level module in a design. Ports are used for communication between modules when Handel-C is not the top-level module in a design.

Few examples exist of programs that target the Xilinx FPGA's and actually run on the Cray XD1. Unfortunately, those examples that do exist are trivial.

A command-line compiler (handelc) runs under Linux to compile Handel-C code targeting a particular FPGA variant. For example, the following command targets a Xilinx Virtex II Pro FPGA on the Cray XD1:

```
handelc -f XilinxVirtexIIPro -p xc2vp50-ff1152-6 -edif \
-I "/usr/share/celoxica/PSL/Hardware/Include" \
-I "/usr/share/celoxica/hardware/include" \
-xl "/usr/share/celoxica/PSL/Hardware/Lib/cray\_xd1.hcl'' \
-D NDEBUG -D __EDIF__ AddOne/main.hcc
```

A version of the Design Kit (DK Design Suite) is available for the Cray XD1. The DK consists of an integrated development environment (IDE) to compile, debug and simulate, and provides a graphical user interface (GUI) in an xterm window (dkdev). To use the GUI tools, some setup is required.

The Platform Developer's Kit (PDK) includes libraries, tools and code to enhance software development for different platforms [8, 26]. The XD1 Platform Support Library (PSL) is a collection of Handel-C interfaces to Cray cores (e.g., rt_core.vhd). Although not supported on the Cray XD1, the PDK was intended to support co-simulation with VHDL, Verilog, SystemC, and MATLAB designs and provides implementations of the PSL, Platform Abstraction Layer (PAL), and Data Stream Manager (DSM) components. PAL provides an application programming interface (API) to enhance portability. DSM provides a data transfer API to enhance communication between the processor (SMP) and the application acceleration processor (FPGA AAP).

Unlike Mitrion-C, Handel-C does not require space on the chip for a virtual processor; whence, more hardware resources are available for the application. To run only a simulation, a license is required; whereas, a free simulator is available to run a simulation based on code developed using Mitrion-C. If the simulation is correct, then the appropriate Handel-C interfaces may be added to run on hardware. ModelSim and Riviera can be used for verification, but these models are not available for the Cray XD1. However, a version of ModelSim that runs under a Microsoft Window OS is available.

The memory map must be defined in a Handel-C file. A programmer configures the FPGA address space following certain conventions for the Xilinx software tools and Cray API. Although these conventions do not depend on the programming language, the various software tools from the different vendors work successfully together because the same conventions are followed.

On the Cray XD1, the FPGA address space occupies a 128 MB region of HyperTransport I/O address space. A programmer divides up this address space into memory ranges using non-overlapping byte addresses. The top 64 MB of FPGA address space may be used for registers. It is convenient to use preprocessor directives to configure the FPGA address space as shown below.

```
#define MEG (1024*1024)
#define TOP ( 128 * MEG )
#define REG_TOP ( TOP - 1 ) // Top of 64 MB region for registers
#define REG_BASE ( 64 * MEG )        // 0x4000000
#define QDR_TOP ( REG_BASE - 1)//Top of 4MB region for QDRII SRAM
#define QDR_BASE (60 * MEG )        // 0x3C00000
#define UNUSED_TOP ( QDR\_BASE - 1 )
#define UNUSED_BASE ( 16 * 1024 )  // 0x4000
#define BRAM_TOP ( UNUSED - 1)//Top of 16 KB region for Block RAM
#define BRAM_BASE ( 0 * MEG )        // 0x0
```

The bottom 64 MB may be mapped to the QDR II SRAMs. In particular, I/O mapped accesses allow a feature called *write combining*. In the source code, the addresses must be shifted three bits to the right to convert byte addresses to quadword (64-bit) addresses.

In addition to the configuration of the FPGA address space, a Handel-C file normally includes calls using macro procedures defined in a PSL library. These macro procedures involve the low-level Handel-C interfaces to VHDL cores (hardware interfaces). On the Cray XD1, typical macro calls include

- `RunRTIf()`

- `RunQDRIf()`

- `RunRTClient( APP_BASE, APP_TOP, AppRead, AppWrite )`

- `RunQDR( QDR_BASE, QDR_TOP )`

- `RunBRam( BRAM_BASE, BRAM_TOP )`

- `RunReg( REG_BASE, REG_TOP )`

`RunRTIf()` sets up the Rapid Transport (RT) core and waits for the RapidArray Processors (RAPs) to come up. `RunRTClient()` runs the RT core using the specified address space and macro procedures. Other calls use RT client and specified macro procedures to use the QDRs, block RAMs and registers.

A programmer must write the C code that runs on the Opteron symmetric multiprocessors (SMPs) using a consistent memory map defined in a Handel-C file, permitting communication between the processor and the coprocessor. Using the Cray API, it is possible to open, close, reset and start the FPGA device, and to read and write registers.

As usual, the C code includes the appropriate header files and is compiled using the gcc command.

After the compiling process is complete, the Xilinx software tools are used to create a bitstream that can be loaded into an FPGA and run. This synthesis process consists of placing and routing the logic on the FPGA, which is performed using a proprietary nondeterministic algorithm. Although compiling is fast, synthesizing can take many hours or days.

4.3.1 CELOXICA EXAMPLE

A Celoxica example (Appendix A) is based on a hyperspectral imaging project headed up by Charles Bachman of NRL's Remote Sensing Division [2]. The project takes multi-spectral images of features such as vegetation, soil and water. For each image, 124 wavelengths are collected. Due to light scattering, much nonlinearity exists in the data, making it difficult to provide an accurate analysis. A non-linear technique, the ISOMAP method [29], based on the observation that the hyperspectral image points appear to lie on a lower dimensional manifold compared to the full 124 dimensional space. The ISOMAP method numerically determines a generalized coordinate system that describes the manifold and the colorization procedure follows from there.

The ISOMAP method is easily applied to small images, but the littoral images usually contain tens of millions of pixels and the resulting calculation easily exceeds available computational time and memory limits of any available computer. The calculation involves the determination of pair wise and minimum path distances between all image points and has cubic scaling. Specialized search algorithms and data structures are used to overcome this computational hurdle. One of the data structures involves a search for intersections between points and point-neighborhoods.

David Gardner (Celoxica) wrote an example as part of Celoxica software support for NRL, addressing the search for intersections. He modified the problem of finding intersections, to finding intersections of points with four-dimensional hypercubes. The point coordinates as well as the hypercube sizes and origins are picked at random using 64 hypercubes and 128K points. The calculation is repeated 500 times so that the wall clock times are on the order of 100 seconds for the Opteron version.

The source code example, including the C code `search.c` and the Handel-C code `search.hcc`, appears in Appendix A. The search is carried out both on the FPGA processor and on the Opteron ASIC processor. The C code demonstrates how to set up the FTR region in order transfer data from the Opteron to the FPGA using the Cray API. The Handel-C code uses the Celoxica macros to handle the data in the FTR region.

The algorithm tested chooses hypercubes such that there is a 50% chance of intersection. The FPGA speed up is dependent on this number – a lower percentage would benefit the software version, a higher percentage benefits the hardware version. Given this caveat, we observe a factor of 113 speed-up on the FPGA compared to the Opteron.

4.4 MITRION-C

Mitrion-C is a novel high-level programming language for the Mitrion Virtual Processor (MVP) [20, 21, 27]. This soft core may be viewed as a virtual machine for which special programming constructs are mapped to custom VHDL code for a variant of an FPGA chip. This virtual machine is a massively parallel computer that permits the type of fine-grain parallelism that characterizes FPGAs and implements speculative evaluation of if-statements. The MVP is not an abstraction and actually runs on an FPGA to carry-out the computations prescribed in the Mitrion-C source code.

Once a programmer learns the Mitrion-C programming paradigm, it is fairly easy to write various programs for practically any supported system and FPGA variant. Presumably, a programmer using a hardware description language (HDL) would spend considerable time writing implementations for new FPGA variants that are being developed at a rapid rate. Yet, Mitrion-C code is hardly portable. The reason is that efficient code is designed to exploit the peculiar architecture. For example, a programmer needs to know how many memory banks are available, their bit-widths, and the I/O operations that are supported. A sound design principle is to first write code in a high-level language such as Mitrion-C and, after testing, design and write HDL code.

Unfortunately, the MVP takes up most of the FPGA resources so that only a fraction of the resources are available to the programmer. The base overhead of the MVP copying inputs to outputs with out any other logic is 11% of the FPGA [60]. Another disadvantage of using Mitrion-C is that the MVP runs at a fixed clock frequency that is slower than current FPGAs.

Mitrion-C replaces conventional arrays in C by so-called "collections." There are three types of these data aggregates: Lists, streams and vectors. While lists and streams are most suitable for the LISP programming model (LISt Processing), vectors are highly suitable for parallel programming. Unlike streams, lists and vectors can be built-up or made smaller using the concatenation (><), take (</) and drop (/<) operators. The operands must be one-dimensional (undocumented), which is not checked by the compiler.

As technology changes rapidly, it is not feasible to fully support every chip unless there is strong market demand. So, it should not be surprising that all features are not supported. Yet, it is amazing that so many different software tools, which are needed to program complex systems, work together as well as they do, at least in the case of Cray, software development by Mitrionics and Xilinx.

Typically, synthesis is possible for Mitrion-C source code that compiles, provided less than half of the potential flip flops are used and memory is sufficiently large. Synthesis is sometimes successful using the standard Xilinx tools even when the number of flip flops is near 75%. Code that produces correct results in simulation and that synthesizes successfully, based on the Xilinx parcheck utility, may either not terminate (i.e., not halt), or produce erroneous results when running on hardware.

In order to synthesize, it may be necessary to reduce the problem size even when only approximately half of the potential flip flops are used. Although the number of statements is important, the size of words in internal memory (Block RAMs) and especially the size of elements in collections will largely limit the size of the problem that can be solved on an FPGA. If an FPGA is used to perform non-trivial operations on a large number of long lists, then it is likely the number of bits used to store the elements is small.

Unlike conventional arrays in C, collections are immutable objects that cannot be modified. Loops in Mitrion-C provide the primary tool to form new collections from existing ones of the same type. Lists and vectors but not streams can be redefined in loops. However, redefining requires that the size is unchanged as the length of lists and vectors is fixed.

Algorithms consisting mainly of loops are easily written in Mitrion-C. The base types for collections are Boolean, floating point and integer. If a programmer has additional information that allows the bit-width of a type to be reduced, then it is advantageous to specify the bit-width to reduce program complexity. In some cases, it is necessary to specify the bit-width, e.g., when shifting left.

A tuple is a small sequence. Unlike collections, tuples may contain different types and instance tokens that are memory states. The base type in memory cannot be a tuple or a collection. Collections may contain tuples, provided all of the tuples have the same sequence of types.

Mitrion-C implicitly executes all instructions in parallel. This programming language lacks a sequential programming construct. By default, Handel-C executes instructions sequentially although the "seq" keyword provides clear documentation. Data dependencies impose an order on computations. Yet, it seems unsatisfactory to never allow sequential computations unless there is a data dependency.

FPGAs provide a memory hierarchy so that internal memories (Block RAMs) offer faster access than external memories. In the Mitrion-C programming environment, all reads and writes take the same amount of time without regard to the type of memory accessed. So, it is not possible to take advantage of the memory hierarchy.

Registers cannot be accessed without performing a read or write operation. A variable may be assigned only once in a scope since there is no concept of order for statement execution. Consider the following sequence of valid statements inside a loop:

```
c = 5; b = c*c;
```

What is the order of execution? What is the value of b? A programmer may be surprised to discover that the answers to the preceding questions depend on whether or not c is a "loop dependent" variable. A loop dependent variable is always explicitly typed and initialized outside a loop. Only loop dependent variables may be returned from a "data-dependent" loop such as a for or while statement. The only variable that may be assigned a new value is loop dependent. A statement such as

```
z = z + 1;
```

is not legal if z is not loop dependent since it is not permissible to assign z more than once in a scope. (What is the value of z on the RHS?) If c is not loop dependent, then the preceding sequence of statements execute sequentially due to the data dependency and the value of b is 25. However, if c is loop dependent and the statements appear in a data-dependent loop, then the statements execute in parallel. For instance, the value of b is 9 if c had the value 3 before the iteration of the loop body.

The for and foreach expressions are loops over a collection. Unlike streams, a list or a vector defined outside a loop, can be referenced in the loop body. In version 1.3, only constants can be used to declare collections inside for and foreach expressions. It might seem that it is not possible to have a variable number of iterations. Functions can be "called" with different constants as Mitrion-C supports polymorphic functions.

To run blocks of code in parallel, a programmer chooses vectors and uses the foreach keyword to unroll the code automatically, i.e., to make copies of the loop body. To implement a loop in a pipeline fashion, use for instead of foreach and iterate over vectors. In practice, a programmer uses only small vectors due to the limited availability of resources to duplicate the body of the loop.

A common technique is to reshape a list into a list of vectors. Then iterate over the smaller vectors using a foreach inner loop. Using this technique, a programmer reduces the resource requirements and still achieves a degree of parallelism. This method is applicable whenever the problem can be decomposed in a suitable way. A good example is matrix multiplication.

An important tool in software development is a simulator with a graphical user interface (GUI). This simulator includes a "throughput analysis" tool that is sometimes useful in identifying bottlenecks in the program.

CHAPTER 5

Case Study: Sorting

Sorting, which is listed as a potential application for acceleration by FPGA coprocessors on the Cray [12], was one of the first, very common, algorithms that we successfully ported to the FPGA. In addition, sorting algorithms are often used in image processing and other scientific applications. The focus of this chapter is sorting on the Cray XD1, using the Mitrion-C and Xilinx tools. In particular, we target a Xilinx Virtex-4 LX FPGA [36].

A list of n elements is optimally sorted in $\Theta(n \log (n))$ time on a sequential computer. There are well-known parallel algorithms for sorting in $O(n \log n)$ time (see for example, [18]). Since an FPGA must initially access every element one at a time to sort a list, the coprocessor requires at least $\Omega(n)$ time to sort a list of n elements; whence, the maximum speedup in sorting a single list of n elements is bounded by $O(\log n)$. Thus, it is reasonable to expect speedup of much less than 100 fold as the number, and length of lists that can be sorted on an FPGA is limited.

The speedup is reduced due to the overhead of loading the logic onto the target device, transferring data (lists) to the local memory of the FPGA, and writing results to the host memory. Unfortunately, the Cray XD1 cannot write to the local memory of the FPGA while the coprocessor is running. To reduce the overhead due to waiting for the FPGA to finish its task, the multiprocessor (SMP) may operate concurrently performing other tasks instead of remaining idle while the FPGA is running.

As FPGAs operate in a regular fashion, the time an FPGA coprocessor takes to perform its task is easily determined. The challenge for the programmer is to keep the multiprocessor as busy as possible doing useful work while the coprocessor is running. Using this approach, speedup is possible even taking into account the relatively large amount of overhead.

Presumably, the overhead will not substantially vary across different implementations as loading the logic, writing by the multiprocessor to the local memory of the FPGA, or by the coprocessor to the host memory occurring at constant rates. We shall focus on comparing different implementations by measuring the time spent while the FPGA is running and ignoring the overhead. The goal is to find out what algorithms and programming techniques yield the best performance based on run time on the Cray XD1. To compare different implementations, define the speedup for each implementation to be the time spent sorting on an AMD Opteron processor using the C standard library function qsort() divided by the time spent sorting on an FPGA coprocessor:

$$\text{speedup} = \frac{\text{time running qsort on SMP}}{\text{time running on FPGA}}.$$

As sorting can be done in various ways, certain requirements are prescribed to conduct a more comparative study. Restrictions are imposed both on the method of sorting and on the type of results returned. Without loss of generality, assume the task is to sort a database based on one of the fields (keys). To access a particular item in sorted order, say the kth item, in another field (column), the original position of the kth key must be known. Hence, it is necessary not only to sort the keys but also to keep track of the original positions.

The original location may be kept by attaching the original location (index) as a prefix to each key in the unsorted list. Define a new data type, say ELEMENT_TYPE, sufficiently large to hold both an index and a key. For example, if the length of a list of 16-bit unsigned integers is 128, then it is convenient to declare

```
#define ELEMENT_TYPE uint:23
```

because seven bits is adequate to hold an index between 0 and 127, and the keys are 16-bit (7+16=23). By shifting an index, say index1, 16 bits to the left

```
ELEMENT_TYPE prefix1 = (ELEMENT_TYPE)index1 << 16;
```

and attaching the "prefix" prefix1 via addition, an "indexed key"

```
r1= prefix1+key1;
```

is obtained. The prefixes (indices) must be ignored whenever comparing keys. Consider two indexed keys, say r1 and r2, with type ELEMENT_TYPE. To check if r1's key is greater than r2's key, write the expression

```
(DATA_TYPE)r1 > (DATA_TYPE)r2
```

where the data type for the keys is declared using a preprocessor directive:

```
#define DATA_TYPE uint:16
```

When the sorting process is complete, only the prefixes are written back to memory, i.e., the keys are ignored. It is convenient to convert the prefixes to have the same type as the keys. To remove the keys and perform this conversion, the expression

```
(DATA_TYPE)(r1 >> 16)
```

where r1 has type ELEMENT_TYPE is used.

The sorting algorithms studied are stable, i.e., two records maintain their relative order when the keys are the same. Although the function qsort() does not actually perform a stable sort, the speedup is nonetheless well-defined.

In each implementation, a coprocessor is required to carry out a stable sort on each list of keys and return a corresponding list whose k^{th} entry yields the original location of the key that belongs in the kth position. So to find the element that belongs in the kth position for a list, look at the index, say i, in the kth position of the list returned. For instance, given the list <7, 7, 5, 2, 6, 4, 1, 3>, the task is to determine the corresponding list <6, 3, 7, 5, 2, 4, 0, 1>, which lists the original positions of the keys in sorted order (counting from zero). To obtain more reliable timing results, the input for each implementation consists of the largest number of lists that fit into the FPGA's external memory (QDRs).

The width of the keys is fixed reasonably small to permit fairly long lists to be sorted. More specifically, every key is a 16-bit nonnegative integer k with $0 \leq k \leq 65534$. The number of lists that be sorted by the coprocessor during a single run depends on the length of each list, which is varied to compare performance using the same algorithm and design techniques.

Although the run time on an FPGA processor does not depend on the input, the time spent sorting on the host processor using `qsort()` does depend on the input. Thus, the speedup will vary depending on the input. To reduce this variability and produce consistent results, a large number of sufficiently varied inputs were used.

The input for each test consists of "pseudorandom permutations" of the list

```
<0, 1, 2, . . . , n-1>
```

where n denotes the length of the list. A simple and fast way to generate pseudorandom permutations is well known. Choose an arbitrary permutation, initially. Successively generate permutations as described next. Uniformly randomly pick any position and swap the element in this position with the last element. After each swap, the rightmost element that was swapped is in its final position. Recursively, uniformly randomly pick a position corresponding to the elements excluding those elements in the tail that have already been placed and swap with the rightmost element excluding the tail. After n-1 swaps, a permutation is generated. This method is fast and produces permutations as if sampling from a uniform distribution [13].

To uniformly randomly pick positions, we employ a multiplicative congruential generator (MCG) compatible with the Cray Mathlib routine RANF [16]:

$$x_{n-1} \ 4485709377909 \ x_n \bmod 2^{48} \ .$$

Although this MCG has a long period, it is a decidedly poor choice [6], but it has worked well for our purposes.

5.1 CONCURRENT SORTS

A common way to improve performance is to execute sorts in parallel. Since the Xilinx Virtex-4 LX FPGAs have four memory banks, it is possible to iteratively perform the following operations in parallel: Read four unsorted lists, sort four lists and write four sorted lists (back to corresponding memory locations). This strategy yields a fourfold improvement in performance. The external memories (QDRs) are dual ported, allowing simultaneous reads and writes.

Another way to improve performance is to carry out I/O operations while doing useful work. Pipelining is a good way to overlap computations with I/O operations. While sorting, it is possible to concurrently both write the previously sorted list and read the next unsorted list.

To some extent, these I/O operations come for free. We describe implementations in which the coprocessor reads unsorted lists from the QDRs and writes sorted lists back to these external memories at corresponding locations. In a typical application, however, the FPGA would write the sorted lists sequentially to the host's memory. Thus, there exists an asymmetry: The coprocessor can concurrently read four lists from the QDRs but can only write sequentially to the host's memory.

To remedy this imbalance, the coprocessor can pack four 16-bit words into a single 64-bit word so that it is possible to effectively write four lists concurrently to the host's memory. This packing is conveniently expressed as a function call

```
w1 = val_4x16to64(x1,x2,x3,x4);
```

where the function `val_4x16to64()` is defined in Figure 5.1. The host processor should also pack

```
val_4x16to64(v0, v1, v2, v3)
{
    bits:16[4] v4x16 = [(bits:16)v0, (bits:16)v1, (bits:16)v2,
(bits:16)v3];
    bits:64 val64 = v4x16;
} val64;
```

Figure 5.1: Pack four 16-bit words into a 64-bit word.

four 16-bit words into a single 64-bit word to efficiently pack more data into the FPGA's external memory. Then a coprocessor unpacks a 64-bit word into four 16-bit words. This unpacking is expediently expressed via a function call

```
(x1,x2,x3,x4) = val_64to4x16( w1 );
```

where the function `val_64to4x16()` is defined in Figure 5.2.

Another strategy to improve performance is to carry out sorts not only in parallel (for each memory bank) but also concurrently (for the same memory bank). Sorting is a relatively slow process relative to the time it takes to read a list. After reading a list and commencing a sorting operation, it is possible to read another list and commence another sorting procedure (provided there are sufficient resources).

Each list in a memory bank can be read sequentially to allow concurrent writes and avoid memory access conflicts. I/O operations may be performed in a round-robin fashion. Each read operation in this round-robin chain is associated with a separate piece of hardware configured to carry out a sorting operation. Thus, many sorts are scheduled in a "circular pipeline." A `foreach` "block expression" is suitable to implement such a circular queue. The programming challenge is to minimize the "idle" time for most of the configured resources.

```
val_64to4x16(v64)

{
    bits:16[4] v4x16 = v64;
    v0 = (DATA_TYPE)v4x16[0];
    v1 = (DATA_TYPE)v4x16[1];
    v2 = (DATA_TYPE)v4x16[2];
    v3 = (DATA_TYPE)v4x16[3];
} (v0, v1, v2, v3);
```

Figure 5.2: Unpack a 64-bit word into four 16-bit words.

5.2   PIPELINING BUBBLESORT

Consider a naive approach to implement a simple brute force algorithm such as bubblesort. A list is sorted after $n-1$ sequential scans of a list of length n. During each successive scan, adjacent elements are compared and swapped if necessary. In this way, elements "bubble" up or down to their final positions. The invariant of the algorithm is that after the kth scan, the kth largest element is placed in its final location.

A traditional programming implementation consists of nested for loops: An outer loop that controls the number of scans (passes) and an inner loop that sequentially performs the swaps for each scan. Unrolling the outer loop creates a pipeline. This means a separate piece of hardware is configured to perform each scan. The input for each such piece of hardware is the output of another piece in the pipeline. This program flow is highly regular, and thus suitable for implementation on an FPGA. Notwithstanding, each successive stage in the pipeline operates relatively faster as the lists become shorter.

Writing all of the results after sorting is completed is impractical because a slow process at the end would become a bottleneck. Due to the invariant of the algorithm, the kth stage may write the kth largest element. Although having multiple stages write results may lead to contention, there is no contention in this case.

A Mitrion-C code fragment that performs a single scan appears in Figure 5.3. The input for a bubblesort pass is a list (s), a memory reference, offset and position. The first for loop extracts the head of the list. The second for loop iterates over a list with one less element (s </ 1). The loop dependent variables are e and Left. No swapping is ever actually done. Instead, the second for loop appends the minimum after every comparison to form a new list (v), which is returned, and keeps track of the largest element seen. The largest element (last) is written at the specified memory location after the loop terminates because of the data dependency.

How can an efficient pipeline be created using the function Bubblesort()? Unfortunately, it is not possible to iterate over a collection (using a for loop) because the lists become shorter in each successive stage. The pipeline is created manually by a sequence of function calls. Each function

```
BubblesortPass( s, Mem0, Start_Offset, Position )
{
    ELEMENT_TYPE Left = 0;
    ELEMENT_TYPE e = for ( a in (s </1) ) { Left = a; } Left;
    (v,last) = for ( enext in (s /< 1) ) {
             (Left, e)  =  if (e > enext) { } (enext,e)
                             else { } (e,enext);
    } (>< Left, e) ;
    Lm = _memwrite( Mem0, Start_Offset + Position, last );
} (v, Lm);
```

Figure 5.3: Function in Mitrion-C performing a single pass of bubblesort.

```
Lm1 = foreach( list_index in < 0 .. N-1 > )
  {
      uint:32 Offset = list_index*LIST_SIZE;
      (v1,Lm1) = BubblesortInitialPass( Mem1, Offset );
      (v2,Lm2) = BubblesortPass( v1, Lm1, Offset, LIST_SIZE-2 );
      (v3,Lm3) = BubblesortPass( v2, Lm2, Offset, LIST_SIZE-3 );
      (v4,Lm4) = BubblesortPass( v3, Lm3, Offset, LIST_SIZE-4 );
      (v5,Lm5) = BubblesortPass( v4, Lm4, Offset, LIST_SIZE-5 );
      (v6,Lm6) = BubblesortPass( v5, Lm5, Offset, LIST_SIZE-6 );
      Lm7 = BubblesortLastPass( v6, Lm6, Offset );
  } Lm7;
```

Figure 5.4: Pipeline in Mitrion-C using function calls.

call creates separate logic that is mapped to the target device. Synchronization is automatic by data dependency.

A sample Mitrion-C code fragment to create a pipeline is given in Figure 5.4. The foreach block iterates over all of the lists in a memory bank (QDR). Each list has eight elements; whence, seven scans are needed. The initial scan takes place while reading the initial unsorted list. The final scan involves two elements which are written to memory after a single comparison. The observed speedup generally increases directly with the length of the list. There are sufficient resources on the chip to sort lists with up to twenty-two 64-bit unsigned integers. The maximum speedup is 9.7 [4].

5.3   A 2-WAY SELECTION SORT IMPLEMENTATION

Selection sort is bubblesort with much less swapping. Each pass of selection sort scans the list for the next largest element and performs a single swap to place that element in its final position. A natural extension is to place the next largest and smallest elements in their final positions after each scan;

whence, half the number of scans are needed. An algorithm for 2-way selection sort is introduced in Algorithm 5.1.

Algorithm 5.1   Algorithm 2-way selection Sort (L).

Input: L   a list
Output: J   indices for sorted list
Requirement: The length of L is even.

---

1. Set S and T to be empty lists and M to be the given list L with each element assigned an index that is the position of the element in L.
2. While M is not empty
   a) Set $m_L = \min(e_1, e_n)$ and $m_R = \max(e_1, e_n)$, where e_1 and e_n denote the first and last elements in M.
   b) Let e_{min} and e_{max} denote the minimum and maximum values in M. In the case of repeated values, choose e_{min} and e_{max} so that the assigned indices are the smallest and largest, and let i_{min} and i_{max} denote the positions in M of e_{min} and e_{max}, respectively.
   c) Append e_{min} to the end of S and insert e_{max} at the beginning of T.
   d) Replace the elements in M at positions i_{min} and i_{max} with the elements m_L and m_R, respectively.
   e) Remove the first and last elements in M.
3. Return the assigned indices J for the list $S\|T$, where $\|$ denotes concatenation.

Example. Apply 2-way selection sort to the list $<0,2,7,4,5,2,1,4>$. For illustration, we label duplicates using subscripts so that a_k indicates this element is the kth occurrence of the element a in the original sequence. The elements of S and T are underlined below. Note, we do not show replacements at the ends (step d) because those elements are removed (step e).

0	2_1	7	4_1	5	2_2	1	4_2
$\underline{0}$	2_1	4_2	4_1	5	2_2	1	$\underline{7}$
$\underline{0}$	$\underline{1}$	4_2	4_1	2_1	2_2	$\underline{5}$	$\underline{7}$
$\underline{0}$	$\underline{1}$	$\underline{2_1}$	4_1	2_2	4_2	$\underline{5}$	$\underline{7}$
$\underline{0}$	$\underline{1}$	$\underline{2_1}$	$\underline{2_2}$	$\underline{4_1}$	$\underline{4_2}$	$\underline{5}$	$\underline{7}$

The first steps of the while loop (steps a and b) permit speculative evaluation. The updates (steps c and d) may be carried out in parallel. The removal of elements in the last step (e) does not incur any cost since these elements may be ignored.

None of the built-in collections seem appropriate to manage the list M. Lists cannot be indexed, which is needed to perform the updates in the 2-way selection sort algorithm. To reduce the amount of resources required, an unrolled `for` or `while` loop is desirable. However, streams cannot be redefined in a loop and vectors require too many resources to perform updates using non-constant indices.

The best data structure to manage the list M is the equivalent of an array stored in internal memory (not QDR). This array is created by calling the function `_memcreate()` to obtain an instance token, say `MemIT`, to perform read and write accesses, as in the following statement

```
DataRAM MemIT = _memcreate( DataRAM  last_IT );
```

where the type declaration for block RAMs is given in the preprocessor directive

```
#define DataRAM mem ELEMENT_TYPE[ L ]
```

in which L denotes the length of a list. These internal block RAMs permit a couple of accesses during each clock cycle. After the last array access, the instance token `last_IT`, passed as an argument to `_memcreate()`, is assigned the last memory reference, (ma), as in the following statement:

```
last_IT = ma;
```

The loops are not unrolled so that there are sufficient resources to perform multiple sorts both concurrently (same bank) and in parallel (four banks). Concurrent sorts may be carried out using function calls to create a pipeline as illustrated in Figure 5.5.

Since five lists cannot be read in parallel from the same external memory bank, a pipeline is created to sequentially read five lists. After the first list has been copied to an internal block RAM, the next list is read from external memory, and so forth, using a chain of function calls to `CopyList()`. Similarly, a chain of function calls to `CopyListBack()` sequentially writes five lists to external memory. After the first list has been sorted, it is copied to external memory. After copying a list to external memory, the next sorted list can be safely copied to the next locations in the same memory bank. Each sort procedure begins as soon as data becomes available.

Implementing the `while` loop in the algorithm using `for` loops in Mitrion-C is tricky because the collection expression for loops must be a constant expression. A solution is to employ a constant stream whose shape matches the control structure of the loop. A code fragment is shown in Figure 5.6.

The inner `for` loop successively iterates 6, 4, 2, and 0 times corresponding to the successive lengths of the stream Pass variable. The dummy variable is not actually accessed in the loop body. The value of the constant Z is not used and its bit width is set to one to minimize resource requirements using a declaration

```
const bits:1 Z = 0;
```

Implementing the `while` loop in the algorithm using a `while` loop in Mitrion-C is given in Figure 5.7. An implementation based on the `while` statement is about twice as fast even though less than half as many sorts are performed on the chip compared to an implementation using only `for` loops [4].

```
(m11,m12,m13,m14,m15,   // similar instance tokens omitted for second,
                        //   third and fourth banks
 l11,l12,l13,l14,l15,   // similar instance tokens omitted for second,
                        //   third and fourth banks
wB1,wB2,wB3,wB4 ) = for( i in (. 0 .. N/M -1 .)  ) {
 /* function calls for bank 1 only */
 /* Copy external memory to block RAM */
 (rd11,smb_11_) = CopyList( rdB1, smb_11, iML );
 (smb_11,slb_11_)=SelectionSort( smb_11_, slb_11 );/* Perform sort  */
 /* Copy block RAM to external memory */
 (slb_11,wr11) = CopyListBack( slb_11_, wrB1, iML );
 (rd12,smb_12_) = CopyList( rd11, smb_12, iML1 ); /* Read next list */
 /* Concurrently perform another sort */
 (smb_12,slb_12_) = SelectionSort( smb_12_, slb_12 );
 (slb_12,wr12) = CopyListBack( slb_12_, wr11, iML1 );
 (rd13,smb_13_) = CopyList( rd12, smb_13, iML2 );
 (smb_13,slb_13_) = SelectionSort( smb_13_, slb_13 );
 (slb_13,wr13) = CopyListBack( slb_13_, wr12, iML2 );
 (rd14,smb_14_) = CopyList( rd13, smb_14, iML3 );
 (smb_14,slb_14_) = SelectionSort( smb_14_, slb_14 );
 (slb_14,wr14) = CopyListBack( slb_14_, wr13, iML3 );
 (rdB1,smb_15_) = CopyList( rd14, smb_15, iML4 );
 (smb_15,slb_15_) = SelectionSort( smb_15_, slb_15 );
 (slb_15,wrB1) = CopyListBack( slb_15_, wr14, iML4 );
 /* similar function calls for bank 2 in parallel omitted;
        returns instance token wrB2 */
 /* similar function calls for bank 3 in parallel omitted;
        returns instance token wrB3 */
 /* similar function calls for bank 4 in parallel omitted;
        returns instance token wrB4 */
 iML=iML+ML; iML1=iML1+ML; iML2=iML2+ML; iML3=iML3+ML; iML4=iML4+ML;
} (smb_11,smb_12,smb_13,smb_14,smb_15, // tokens omitted for second,
                                       //   third and fourth banks
   slb_11,slb_12,slb_13,slb_14,slb_15, // tokens omitted for second,
                                       //   third and fourth banks
   wrB1,wrB2,wrB3,wrB4);
```

Figure 5.5: Sorting concurrently in a pipeline.

5.4   PARALLEL BUBBLESORT

A parallel version of bubblesort is called an "odd-even sort." Denote a list L with n elements by

$$L = <a_0, a_1, \ldots, a_p, a_{p+1}, \ldots, a_{n-1}> .$$

```
Two_Way_Selection_Sort( u )
{       // Initialization code (omitted)
    INDEX_TYPE Start=0; INDEX_TYPE End = L_1;
    (ma,mb) = for( STREAM_BASETYPE (..) Pass in (.
        (.Z,Z,Z,Z,Z,Z.),
        (.Z,Z,Z,Z.),
        (.Z,Z.),
        (.-.) .) )
        {       // Initialization code (omitted)
            INDEX_TYPE j=Start+1;
            (i_Min,i_Max,e_Min_i,e_Max_i, lm)=for( dummy in Pass )
            {       // body (omitted)
                j = j + 1;
            } (iMin, iMax, eMin_i,eMax_i, memA);
            // updates (omitted)
            Start=Start+1; End = End - 1;
        } (mema, memb );
        // Termination code (omitted)
} (ma,mb);
```

Figure 5.6: A 2-way selection sort function in Mitrion-C using an inner for loop.

Any pair of consecutive elements, say (a_p, a_{p+1}), is said to be even or odd if the position p of the first element of the pair is even or odd, respectively, e.g., (a_0, a_1) is an even pair whereas (a_1, a_2) is an odd pair.

Algorithm 5.2   Algorithm parallel `Bublesort(L)`.

Input:	L	a list with n elements
Output:	J	indices for sorted list
Requirement:	The length of L is even.	

1. Sort all even pairs in parallel.
2. Sort all odd pairs in parallel.
3. Repeat steps 1-2 sequentially exactly $\frac{n}{2}$ times.

Example. Apply parallel bubblesort to the list $<0,2,7,4,5,2,1,4>$.

```
Two_Way_Selection_Sort( u )
{        // Initialization code (omitted)
    INDEX_TYPE Start=0; INDEX_TYPE End = L_1;
    (s,t,ma) = for( INDEX_TYPE Pass in < 1 .. MIDR > )
            {        // Initialization code (omitted)
               INDEX_TYPE j = Pass;
               (i_Min,i_Max,e_Min_i, e_Max_i, Lm) =
                               while(j < End) iterations 1
                {        // body (omitted)
                    j = j + 1;
                } (iMin, iMax, eMin_i, eMax_i, memA);
                // updates (omitted)
                Start=Start+1; End = End - 1;
            } ( >< e_Min_i, >< e_Max_i, mema);
        // Termination code (omitted)
} (s,t);
```

Figure 5.7: A 2-way selection sort function in Mitrion-C using an inner while loop.

List:	0	2_1	7	4_1	5	2_2	1	4_2
even:	0	2_1	4_1	7	2_2	5	1	4_2
odd:	0	2_1	4_1	2_2	7	1	5	4_2
even:	0	2_1	2_2	4_1	1	7	4_2	5
odd:	0	2_1	2_2	4_1	1	7	4_2	5
even:	0	2_1	2_2	4_1	1	7	4_2	5
odd:	0	2_1	2_2	1	4_1	4_2	7	5
even:	0	2_1	1	2_2	4_1	4_2	5	7
odd:	0	1	2_1	2_2	4_1	4_2	5	7

A vector is the only collection that permits parallel access. Unlike the even stages, the odd stages do not involve all elements. By partitioning lists into two vectors containing the even and odd elements, it is possible to iterate over pairs of vectors in a foreach loop to access all elements in parallel. The lists are formed when the unsorted lists are read from external memory and reformatted as vectors. The elements that are not used during the odd stages must be dropped and added back. Figure 5.8 shows a Mitrion-C code fragment implementing parallel bubblesort for a list with eight elements. The maximum speedup sorting lists with up to forty 64-bit unsigned integers is 30 [5].

5.5   COUNTING SORT

Counting sort determines the position of an element by counting the number of smaller elements. For example, the minimum and maximum appear in the first and last positions, respectively. The count is unique whenever all elements are distinct. To force the algorithm to be stable in the case

```
(e1,o1,e2,o2,e3,o3,e4,o4) =
   foreach( ELEMENT_TYPE Index1 in <0,2,4,6> ) {
      // code omitted: read an even element from each bank,
      //   attach prefixes and label the results ai,ci,ei,gi.
      // code omitted: read an odd element from each bank,
      //   attach prefixes and label the results bi,di,fi,hi.
} (ai,bi,ci,di,ei,fi,gi,hi);

ELEMENT_TYPE[4] EvenVector = reformat ( e, [4] );
ELEMENT_TYPE[4]  OddVector = reformat ( o, [4] );
(x, w) =  for ( EvenOddPasses  in  <1..4> ) {
    /** Pass: even **/
    (EvensEP, OddsEP) =
         foreach ( a,b  in  EvenVector, OddVector ) {
              (p,q) = if ( a > b ) (b,a) else (a,b);
      } (p,q);

    /** Pass: odd **/
    (OddsOP,EvensOP) =
         foreach ( a,b  in  (EvensEP /< 1),(OddsEP </ 3)) {
              (p,q) = if ( b > a ) (a,b) else (b,a);
      } (p,q);

    /* Add first and last elements back **/
    (EvenVector, OddVector) =
         foreach ( a,b in ( EvensEP </ 1 ) >< EvensOP,
                              OddsOP >< (OddsEP /< 3 ) )
      {} (a,b);

 } (EvenVector, OddVector);
```

Figure 5.8: Parallel bubblesort code fragment in Mitrion-C.

of duplicates, the count includes the number of identical elements that appear earlier in the original list.

Algorithm 5.3   Algorithm counting sort(L).

Input:	L a list
Output:	J indices for sorted list

1. For each position in the list L, count the total number of elements less than the element appearing at this position plus the number of identical elements appearing before this position in L.
2. Return the sums.

Example. Apply counting sort to the list $<0,2,9,4,5,2,1,4>$. By inspection, the counts are $0,2,7,4,6,3,1$ and 5. This is done by counting the list elements in ascending value beginning at 0, as shown in the following diagram:

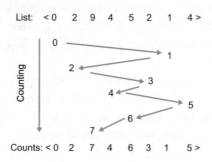

```
CountingSort ( Mem1, Offset )
{
  list = foreach ( Index in <0.. L_1 > )  {
       SHORT_TYPE data =_memread ( Mem1, Offset+Index );
  } data;

  v = reformat( list, [ L ] );

  Last_m = foreach ( e  in  v  by  INDEX_TYPE  Pos_e )  {
      INDEX_TYPE sum = 0;
      Sum = for ( Pos_a , a  in  <0.. L_1> , list  ) {
        sum = if ( a > e ) {
               } sum else {
                   le= if ( (a==e) && (Pos_a >= Pos_e) )  {
                       } sum else {
                         s = sum + 1;
                       } s;
               } le;
      } sum;
      m  = _memwrite ( Mem1, Offset + Sum,  Pos_e );
  } m;
} Last_m;
```

Figure 5.9: Counting sort function in Mitrion-C.

Figure 5.9 displays a counting sort function in Mitrion-C. Sequential counts are performed in parallel in a foreach loop over a vector. The counts are accumulated by iteration over a list in a for loop. The maximum speedup for lists with up to 50 64-bit elements is 32 [5].

5.6   SIMULATION

Before building and running code on hardware, a simulation is run to check correctness. If the simulation is not correct, then fix the errors, possibly redesign and rewrite part of the code. Then rerun the simulation. If the simulation is correct, then the code still might not run as expected on hardware due to bugs in the software tools, which should be reported. For example, implementations for matrix multiplication ran correctly only in simulation. The bug was reported and fixed in the most recent release of the software.

The simulator is not clock accurate. The number of steps reported in batch and graphical user interface (GUI) modes is different because additional optimizations are performed in batch mode. A programmer might modify a code so that it runs faster during simulation. However, a programmer may observe that such improved code runs slower on hardware. The only valid measure of performance is the run time on hardware.

A free simulator is available. This simulator displays a data dependency graph and optionally shows names and types. There is a natural correspondence between the data dependency graph and the program constructs. The advantage of using this simulator on a personal computer is that the option of visualization (GUI mode) is useful. Using the GUI mode when running the simulator remotely on the Cray XD1 is not practical as visualization runs slowly over the network. Using the free simulator it is not possible to specify a platform, simulate the FTR region, or handle complex types (such as packed data).

Input from external memory banks for a simulation may be automatically read from specified ASCII files, either created beforehand, or seemingly generated randomly on-the-fly. For instance, to compile and simulate a source file called source.mitc in GUI mode targeting a Virtex II FPGA, enter the following commands:

```
mitrion   -gui -jvm "-Xms512m -Xmx2400m" -sizes \
          -platform xd1vp50 -output out \
          -input b1 b2 b3 b4 -- source.mitc
```

where b1, b2, b3 and b4 are input files corresponding to the RAM banks, and out is the prefix for the corresponding output files. To run in batch mode, replace the -gui flag by the -batch flag. The number of input files must match the number of arguments to function main(). One method of creating input files is to develop a simple code to create the input and run a simulation to output the data to ASCII files. Stopping and resetting a simulation does not reopen files, which is a bug.

The Throughput Analysis window in GUI mode that helped to spot bottlenecks in the design is no longer available as of release 1.5.1. To print estimates of performance statistics when compiling code on the Cray XD1, add the switch -log SUMMARY on the command-line. For example, compile the source code source.mitc targeting a Virtex-4 FPGA use the command:

```
mitrion   -jvm "-Xms512m -Xmx2400m" -sizes -gen-c-header \
          -log SUMMARY -configure platform xd1lx160-m2 \
          source.mitc
```

5.7   MEMORY CONVENTIONS AND INTERFACES

Although there is a greater flexibility using the free simulator, there are precise restrictions on the number and order of arguments and return types of the function main() when targeting an FPGA. There must be four or more arguments and return values, each of which is a 64-bit memory reference. The kth memory reference corresponds to the kth external memory bank (QDR), k =1,2,3,4. An optional fifth reference corresponds to the host memory (FTR transfer region). Up to 32 additional 64-bit memory references correspond to registers, accessible by the host and coprocessor. The coprocessor can only read values (input registers) written by the host processor before starting the FPGA. The host can only read values (output registers) that the coprocessor writes after it finishes its task.

A typical main() function declaration is given in Figure 5.10. The size of each external (QDR)

```
// Type declaration for a register:
#define Register uint:64
// Type declaration for FPGA external memory:
#define ExtRAM mem bits:64[0x80000]
// Type declaration of HOST memory
#define HostRAM mem uint:64 [0x2000000000]

(ExtRAM, ExtRAM, ExtRAM, ExtRAM, HostRAM, Register)
 main
(ExtRAM m_a, ExtRAM m_b, ExtRAM m_c, ExtRAM mem_d, HostRAM
 Mem_Host, Register ftr_byteaddr, Register ftr_offset )
{
  Register ftr_array = ( ftr_byteaddr >> 3 ) + ftr_offset;
  /* code omitted;
     returns 64-bit value in last_register, plus last
     memory references m_B1, m_B2, m_B3, m_B4, and
     LMem_Host */
  Register LReg = last_register;
} ( m_B1, m_B2, m_B3, m_B4, LMem_Host, LReg );
```

Figure 5.10: Typical main() function in Mitrion-C.

memory bank is specified as 0x80000 (0x100000 for release 1.4.1 or earlier) 64 bit words (4 MB). The size of the FTR transfer region must be specified as 0x2000000000 64 bit words (this value is required by the Mitrion-C compiler). The host processor writes the FTR address and offset into the input registers at offset 0x40 and 0x41. Additional input registers are written at offsets 0x42-0x5F.

The coprocessor reads the byte address and converts it into a quadword address and adds the base offset. The host processor accesses the output registers (such as LReg) at offsets 0x60-0x7F.

The Cray API can be used to interface with the Mitrion Virtual Processor (MVP). The starting offset for the first 4 MB memory bank is 64 MB (0x4000000 bytes). The second, third and fourth banks are mapped to contiguous addresses immediately following the end of the first bank. Sample code using the Cray API is illustrated in Figure 5.11.

Whenever the status is returned, errors may be handled using a statement as follows

```
if (status < 0) {/* handle error */ }
```

not shown in Figure 5.11 to simplify the presentation. The Mitrion host abstraction layer (Mithal) API provides another interface to the MVP. The header (mithal_gen.hdr) is generated while compiling Mitrion-C code. Although there is no compelling reason to use the Mithal API instead of the Cray API, the reference document for the former API is useful because the latter API is not fully implemented for the MVP [30]. Compile C code (which uses FPGAs), say source.c, using a command line feed as follows

```
gcc   -o source -O3 -Wall -m64 -D_REENTRANT
      -I/share/mitrion/mithal/include -I/usr/local/include \
      -D_GNU_SOURCE source.o -L/usr/local/lib64 -lufp -lpthread -lrapl
```

5.8   GENERATING LOGIC AND RUNNING CODES

To synthesize, load the appropriate modules (mitrion and xilinx), copy the VHDL file (user_app.vhd), generated while compiling Mitrion-C code, to a "user_app" directory, and change to the "place and route" directory. A bitstream that can be loaded onto a chip can then be created by using a makefile and the commands:

```
make clean
make dist_clean
make top.bin
```

To check a bitstream after synthesis, run:

```
parcheck -nogui ./
```

in the scratch directory containing the generated files. Note that a program may run correctly even when the parcheck utility reports errors.

On the Cray XD1, to run an executable on hardware submit a PBS batch file in the subdirectory of scratch containing the executable and bitstream files. To run an executable, fpga_app, on a compute node with a Virtex II Pro FPGA write script file, fpga_app.pbs shown in Figure 5.12, and submit the job from the directory containing the executable

```
qsub fpga_app.pbs
```

```c
#define    __USE_XOPEN2
#define _XOPEN_SOURCE 600
#include "ufplib.h" //add headers fcntl.h,argp.h,einlib.h, unistd.h
#define MEM_SIZE        (8 * 1024 *1024)
#define MEM_DISTANCE    (8 * 1024 *1024)
#define MEM_OFFSET      (64 * 1024 *1024)
#define ARRAY_SIZE      33423360              /* Size of array */
#define TYPE_VALUE      0x0UL
#define TYPE_ADDRESS    0x1UL
typedef  unsigned long  u_64; typedef  uint16_t  Element_Type;
Element_Type * Bank[ 4 ];      void * ftr_mem;
int main (int argc, char** argv) {
  int fpga_id,status,num_bytes,flags=O_RDWR|O_SYNC,
      mmap_flags=PROT_READ | PROT_WRITE;
  err_e err;  size_t length = MEM_SIZE; off_t Offset = MEM_OFFSET;
  const char *loadfile = "top.bin.ufp";
  fpga_id = fpga_open ("/dev/ufp0", flags, &err); // Open FPGA
  if (fpga_id < 0) {/* handle error */ }
  // Load logic into FPGA (can also be loaded in PBS batch job):
  num_bytes = fpga_load(fpga_id, loadfile, &err);
  if (num_bytes < 0) { /* handle error */ }
  // Map QDRs into application address space:
  for (i=0; i < 4; i++) { Bank[i] = (Element_Type*)fpga_memmap(fpga_id,
      length, mmap_flags, MAP_SHARED, Offset + i*MEM_DISTANCE, &err);
    if (Bank[i]==NULL){/* handle error */}}
  // Setup host memory for FPGA use:
  size_t pagesize = getpagesize();
  size_t pages_needed  = ( ARRAY_SIZE*sizeof (u_64) - 1)/ pagesize + 1;
  size_t buf_size = pages_needed * pagesize;
  if
  int code = posix_memalign ( &ftr_mem, pagesize, buf_size );
  if (code != 0) { /* handle error */ }
  // Register FTR region:
  fpga_register_ftrmem(fpga_id, ftr_mem, buf_size, &err);
  if (err != NOERR) {/* handle error */ }
  status = fpga_reset(fpga_id, &err);   // Reset logic on FPGA
  status = fpga_start(fpga_id, &err);   // Release the logic from reset
  status = fpga_wrt_appif_val ( fpga_id, 0x00000000000000001UL, (0x01*
          sizeof (u_64)) + 0, TYPE_VALUE, &err); // Stop FPGA
  // Provide FTR address via first input register:
  status = fpga_wrt_appif_val ( fpga_id, (u_64)ftr_mem,
          (0x40* sizeof (u_64)), TYPE_ADDRESS, &err);
  // Provide FTR offset via second input register:
  status = fpga_wrt_appif_val(fpga_id, ftr_offset,
          (0x41* sizeof (u_64)),TYPE_VALUE, &err);
  // Clear processor state before running
  status = fpga_wrt_appif_val ( fpga_id, 0x0000000000000000UL,
          (0x02* sizeof (u_64)), TYPE_VALUE,  &err);
  status = fpga_wrt_appif_val ( fpga_id, 0x0000000000000000UL,
          (0x01 * sizeof (u_64)), TYPE_VALUE, &err); // Start the AAP
  do {status = fpga_rd_appif val ( fpga_id, (void*)&Value,
                (0x02*sizeof(u_64)), &err);}  // Wait for FPGA
  return 0; }
```

Figure 5.11: C code using Cray API.

```
!/bin/ksh
#PBS -l nodes=1:ppn=4#excl
#PBS -j oe
#PBS -l walltime=0:30:00
#PBS -V
cd $PBS_O_WORKDIR
./fpga_app
```

Figure 5.12: Sample PBS script.

If a job is running a long time compared to the time it would take the host processors to accomplish the same task without using FPGAs, the job should be killed (qdel -Wforce job_number). In such cases, there may be a bug in the compiler. If a job crashes a node, an attempt has been made to use memory in some unsupported way.

CHAPTER 6

Alternative Technologies and Concluding Remarks

6.1 ALTERNATIVE TECHNOLOGIES

Other high-level languages to program FPGAs include Dime-C [50] from Nallatech and Impulse-C [49] from Impulse Accelerated Technologies. Among the FPGA languages currently available, some are easier to learn than others. Notwithstanding, none of them completely remove the various hurdles a programmer must confront.

The newest entry into the reconfigurable computing arena by Convey Computer Corporation is the HC-1, containing four Xeon CPUs with access to four Virtex-5 FPGA coprocessors [9]. The shared memory model integrates the CPU and the FPGA, eliminating some of the overhead of using FPGA technology. Programming the HC-1 involves using precompiled code and sophisticated compilers to configure the target device. This means a programmer becomes simply a user of the technology and does not write FPGA code and wait hours to synthesize a bit file that would later be loaded to configure an FPGA device. Instead, a programmer learns special directives that are wrapped around the code segments to be executed on the FPGA. Both FORTRAN and C are supported.

How an FPGA on the HC-1 can be configured depends entirely on the "personality" – a set of packages of the precompiled parts. Currently, only a small number of personalities are available. Hence, programming is severely limited by the choice of a personality, which may be adequate for only some applications. To achieve exceptionally good performance, a personality likely provides more specialized as opposed to more general functionality. Casual programmers cannot develop a personality. For example, a team of engineers at Convey and scientists at the University of California San Diego developed a "Proteomics Search Personality" which yielded a 100-fold speedup in performance. The Proteomics Search algorithm conducts an unrestricted, blind search of a massive protein database to look for protein structure modifications. Such a search was not possible prior to the HC-1 [9].

Due to Moore's Law for FPGAs, [46] the number of resources and capabilities of FPGAs continue to steadily increase. Nevertheless, HPC still presents a significant design challenge for FPGAs. Larger functional blocks are required to handle the greater computational density found in supercomputer applications. Larger memory units are required to handle double precision values ubiquitous in HPC calculations. Faster clock speeds are required to keep pace with the CPU speeds.

Finally, a faster interface between the CPU and FPGA is required – tighter connectivity with the CPU, faster reconfiguration times and more I/O ports for streaming data to and from the host CPU.

More generally, an FPGA is just one type of hardware accelerator currently available. PACT XPP technology is based on an array of units, which may be regarded as specialized FPGAs. Multicores, GPUs, Cell processors, array processors and other hardware accelerators are being offered to the HPC community in effort to provide faster computational platforms. The motivation for this diverse spectrum of accelerators is two fold – increased computational speed and decreased power consumption [51]. Yet, the costs for the hardware components and the software licenses cannot be ignored [60].

Multicore systems lie at one end of the spectrum of hardware accelerators. These systems not only provide increased speed but also reduced power consumption per core since less circuitry is needed than in a traditional multi-CPU single-core platform. The disadvantage is that the cores located on the same node of a multicore system compete for the limited node memory bandwidth, which can limit speedups. On the other hand, there are no software hurdles; the programming model is identical to that of the single core multi-CPU system.

At the other end of the spectrum are FPGAs. FPGAs provide a significant reduction in power consumption, which is just a fraction of a traditional CPU [34]. With the potential massive parallelism provided by FPGAs, speedups can be on the order of what is equivalent to 100s or even 1000s of cores. The limitation here is the programming effort. Not only is the programming effort a major obstacle, but licenses for these high-level languages can be on the order of thousands of dollars for a single seat.

One may consider graphics processing units (GPUs) to lie in the middle of the spectrum between the two extremes of multicore systems and FPGAs [61]. GPUs are currently the most popular hardware accelerator. Although the total power consumption is typically double that of a traditional CPU (power savings is evident in terms of power consumption per GFLOPS) and the programming effort substantial, the cost is much reduced since it is based on commodity graphics cards and, hence, the concomitant economies of scale [34].

Nvidia has recently come out with its Tesla system [40]. It is the first graphics system to target the HPC community. The system combines two quad core Xeon's with four GPUs consisting of 240 cores. Programming is supported with CUDA, a language like C, with massively parallel constructs.

ATI, now part of AMD, has come out with its FireStream product and Brook+ programming language while Intel is coming out with its own GPU, codenamed Larabee that will support the x86 instruction set [1, 37]. All of these systems have native support for IEEE double precision arithmetic reflecting a strong focus on scientific programming. On the other hand, these GPU programming languages are based on the stream-processing model. This model is most efficient when computations can be cast within a SIMD algorithm. In this sense, FPGAs are more flexible computational engines.

Even without hardware accelerators, the computational landscape is changing. Today's top HPC performing systems are on the order of hundreds of thousands of cores. Such systems are not only expensive to power but also to program. MPI codes that run without error on 128 cores often hang or crash on a multi-thousand core system [52]. Newer programming languages, like Titanium, Co-Array FORTRAN, and Unified Parallel C, are examples of PGAS (Partitioned Global Address Space) languages that hope to ease the cost of program development and take advantage of the latest hardware [38]. Also, MPI libraries, such as PETSc, are being employed to obviate explicit MPI programming [39].

6.2   CONCLUDING REMARKS

FPGA acceleration is a new technology that promises dramatic increases in computational performance for some applications. Because of their unique technological implementation, these devices aptly combine software and hardware into a reconfigurable architecture. The strongest advantage of using FPGAs is their ability to exploit data parallelism, instruction concurrency, and pipelining.

Today, reconfigurable supercomputing is a reality, and major manufacturers of high performance computers are currently building machines with advanced FPGA acceleration units. Furthermore, some preliminary experiments have shown that reconfigurable supercomputing offers increased performance of several orders of magnitude for selected examples. Unfortunately, to date, there is no easy way to accelerate a traditional parallel code using FPGAs. There is no magical compiler flag that generates ready to use reconfigurable code. Instead, programmers have the entire burden of deciding what tasks to assign to the coprocessor and when to start and stop the device, in addition to learning new programming tools and new programming paradigms. Ultimately, users must redesign and rewrite major portions of code. As such, reconfigurable supercomputing is complex and a time consuming endeavor.

To advance the use of FPGAs, it is not enough to increase the performance and capabilities of the commodity components; integrated systems are needed. For example, more I/O ports are needed on FPGA device to advance parallel computations, especially systolic computations, to more fully utilize the device capabilities. Shared memory is an important integration to eliminate the overhead of communication costs. In particular, a system of a large array of FPGAs with relatively few CPUs is promising to advance fine-grain parallelism.

Sophisticated compilers are also needed to shift the burden of scheduling tasks to run on the coprocessor. Using profiling and tracing, a programmer decides what code runs on which (co-)processor, writes all of the code, and the compiler performs scheduling and reconfiguration of the device as prescribed by the program. Shifting and rearranging the code to meet timing requirements is a task better suited to a compiler.

Today, the HPC industry seems to be at the cross roads where long held traditions of both hardware and software are being challenged. It is highly likely that the changes observed in supercomputing over the past couple decades will pale compared to the changes on the horizon in the next couple decades.

APPENDIX A

search.c

```c
#include "einlib.h"
#include <stdio.h>
#include <stdlib.h>
#include <fcntl.h>
#include <argp.h>
#include <string.h>
#include <unistd.h>
#include <time.h>
#include <math.h>


/* Configuration Parameters */
/* Note, the probability is per dimension.
   So, for a 4-D hypercube, the probility of point-hypercube
   intersection is PROB^4.  0.841^4 = 50% */
#define PROB .841
#define NBR_DIMS 4
#define NBR_HYPERCUBES 64
//#define NBR_POINTS 1000000
#define NBR_POINTS 256 * 1024
#define VERBOSE_DEPTH 10
#define REPEAT_COUNT 500



/* Derived Configuration Parameters */
#define RESULT_FIELDS (((NBR_HYPERCUBES-1)/64)+1)

/* Define the FPGA write types (address or data value) */
#define TYPE_VAL  0x0UL
#define TYPE_ADDR 0x1UL

/* Define the addresses of the FPGA Registers */
#define MEGA 1024 * 1024
#define XY_REG          ( 0 * MEGA)
```

```c
#define RESULT_REG         ( 0 * MEGA)
#define RECT_CFG_REG       (64 * MEGA)
#define RESULT_PTR_REG     (64 * MEGA + 0x400UL)

/* C++ wannabe */
typedef char bool;
#define true 1
#define false 0

/* Declare types for unsigned integers. */
typedef unsigned long  u_64;
typedef unsigned short u_16;

typedef struct {
    u_16 d[NBR_DIMS];
} point;

typedef struct {
    u_16 min;
    u_16 max;
} dim;

typedef struct {
    dim d[NBR_DIMS];
} hypercube;

typedef struct {
    u_64 f[RESULT_FIELDS];
} result;

/****************************************************************************/
/* Commmand line parsing                                                    */
/****************************************************************************/
static struct argp_option options[] = {
  {"verbose", 'v', 0, 0, "Produce verbose output"},
  {0}
};
static error_t parse_opt (int key, char *arg, struct argp_state *state);
```

```
struct arguments
{
    int  verbose;   /* The -v flag */
};

static char args_doc[] = "";
static char doc[] = "Demonstration of a Celoxica design.";
static struct argp argp = {options, parse_opt, args_doc, doc};

/* Provide a function to parse the parameters. */
static error_t parse_opt (int key, char *arg, struct argp_state *state)
{
    struct arguments *arguments = state->input;

    switch (key) {
    case 'v':
        arguments->verbose = 1;
        break;
    case ARGP_KEY_ARG: // no arguments accepted
        if (state->arg_num >= 0) {
        argp_usage(state);
        }
        break;
    case ARGP_KEY_END:
        break;
    default:
        return ARGP_ERR_UNKNOWN;
    }

    return 0;
}

int print_err (err_e e)
{
    switch (e) {
    case NOERR:
        printf("Success.\n");
        break;
```

```c
        case FILEOPRERR:
            printf("File operation system call failed.\n");
            break;
        case INVALIDOP:
            printf("Invalid API operation requested.\n");
            break;
        case INVALIDVAL:
            printf("Invalid value passed to the API call.\n");
            break;
        case INVALIDARGS:
            printf("Invalid argument passed to the API call.\n");
            break;
        case INVALIDINP:
            printf("Invalid input given to the API call.\n");
            break;
        case DEVOPRERR:
            printf("FPGA device operation error.\n");
            break;
        case UNKNOWNERR:
            printf("Unknown error.\n");
            break;
        default:
            break;
        }
        return 0;
}



/****************************************************************************/
/* init_data_structures                                                     */
/* populate the input data structures with points & hypercubes              */
/****************************************************************************/
int init_data_structures(struct arguments *arguments, point *points,
                         hypercube *hypercubes) {
    u_16 d1, d2, width, start_max;
    int i,j;

    for( i = 0; i < NBR_POINTS; i ++ ) {
        for( j = 0; j < NBR_DIMS; j ++ ) {
```

```
            d1 = (u_16)random();
            points[i].d[j] = d1;
        }
    }


    width = (-1);
    width = width * PROB;
    start_max = (-1 - width);
    if (arguments->verbose) {
    printf("    start_max=%08x  width=%08x \n", start_max, width);
    }
    for( i = 0; i < NBR_HYPERCUBES; i ++ ) {
        for( j = 0; j < NBR_DIMS; j ++ ) {
            d1 = (u_16)( random() % (start_max+1));
            d2 = d1 + width;
            hypercubes[i].d[j].min = d1;
            hypercubes[i].d[j].max = d2;
        }
    }


    if (arguments->verbose) {
        printf ("Displaying first %i points and hypercubes:\n",
                VERBOSE_DEPTH);
        point * p = points;
        for( i = 0; i < VERBOSE_DEPTH; i ++ ) {
            printf(  "    point[%i]\n",i );
            for( j = 0; j < NBR_DIMS; j ++ ) {
                printf(  "        dim[%i] %04X\n", j, (*p).d[j] );
            }
            p ++;
        }

        hypercube *s = hypercubes;
        for( i = 0; i < VERBOSE_DEPTH; i ++ ) {
            printf(  "    hypercube[%i] \n", i );
            for( j = 0; j < NBR_DIMS; j ++ ) {
                printf ("        dim[%i] {%04X:%04X}\n",
                        j, (*s).d[j].min,(*s).d[j].max );
            }
```

```
                s++;
            }
        }

    return(0);
}


int search_hw(struct arguments *arguments, int fpga_id, point *points,
              hypercube *hypercubes, result **resultsPtr) {
    err_e e;
    int i, j;

    /* Allocate memory where FPGA can store results */
    result *results_temp;
    results_temp = fpga_set_ftrmem(fpga_id, 9, &e);
    *resultsPtr = results_temp;

    /* Declare a pointer to the "hypercube" memory in FPGA */
    int hc_mem_size = NBR_HYPERCUBES * sizeof(hypercube);
    u_64 *hcdst = fpga_memmap(fpga_id,hc_mem_size, PROT_READ |
                            PROT_WRITE, MAP_SHARED, RECT_CFG_REG, &e);

    /* Declare a pointer to the "point" memory in FPGA */
    int pnt_mem_size = 64 * 1024 * 1024;
    u_64 *pointdst = fpga_memmap(fpga_id, pnt_mem_size, PROT_READ |
                            PROT_WRITE, MAP_SHARED, XY_REG, &e);

    /* Load the hypercubes to be searched into the FPGA */
    printf("    Loading hypercubes into FPGA\n");
    memcpy(hcdst, (u_64 *)hypercubes, hc_mem_size);

    /* Optional test to make sure that the hypercubes are loaded */
    if (arguments->verbose) {
    u_64 read_data, expected;
    fpga_rd_appif_val (fpga_id, &read_data, RECT_CFG_REG, &e);
    expected = *((u_64*)(hypercubes));
    if ( read_data != expected ) {
        printf("    FAILED TO LOAD FPGA.  Expected: %0161X  Received: "
               "%0161X\n", expected, read_data);
```

```
            return(1);
        } else {
            printf("    Hypercubes loaded successfully\n");
        }
    }

    /* Search for intersects - send each point to the FPGA (the FPGA will
       write results back into processor memory) */
    printf("    Searching\n");
    for( i = 0; i < REPEAT_COUNT; i ++ ) {
        // tell the FPGA where to store the results
        fpga_wrt_appif_val(fpga_id, (long)results_temp, RESULT_PTR_REG,
                           TYPE_ADDR, &e);
        memcpy(pointdst, (u_64*)points, NBR_POINTS * sizeof(point));
    }

    // this will force the FPGA to flush any remaining results
    fpga_wrt_appif_val(fpga_id, 0x0, RESULT_PTR_REG, TYPE_VAL, &e);
    printf("    Finished\n");

    if (arguments->verbose) {
        printf ("    Displaying first %i results:\n", VERBOSE_DEPTH);
        result * r = results_temp;
        for( i = 0; i < VERBOSE_DEPTH; i ++ ) {
            printf( "        point[%i]\n",i );
            for( j = 0; j < RESULT_FIELDS; j ++ ) {
                printf( "           result[%i] %016lX\n", j, (*r).f[j] );
            }
            r ++;
        }
    }
    return(0);
}


int search_sw(struct arguments *arguments, point *points,
              hypercube *hypercubes, result **resultsPtr) {
    result *results;
    results = malloc( NBR_POINTS * sizeof(result));
```

```c
*resultsPtr = results;

int rc, i, j, k;
u_16 pnt, min, max;
bool hit;

point *p;
result *r;
hypercube *s;

printf("    Searching\n");
for( rc = 0; rc < REPEAT_COUNT; rc ++ ) {
p = points;
r = results;
for( i = 0; i < NBR_POINTS; i++, p++, r++ ) {
    s = hypercubes;
    (*(u_64*)(r)) = 0; // clear memory
    for( j = 0; j < NBR_HYPERCUBES; j++, s++ ) {
    hit = true;
    for( k = 0; k < NBR_DIMS; k ++ ) {
        pnt = (*p).d[k];
        min = (*s).d[k].min;
        max = (*s).d[k].max;
        if( (pnt < min ) || (max < pnt) ) {
        //printf( "><> Point %i is outside of hypercube %i on "
        //        "dimension %i\n", i, j, k );
        hit = false;
        break; // don't bother checking the remaining dimensions
               // of this hypercube
        }
    }
    // update results
    if( hit ) {
        //printf( "><> Yippie! Point %i intersects hypercube"
        //         "%i\n", i, j );
        (*r).f[j/64] += ((u_64)1) << (j % 64);
    }
    }
}
```

```
    }
    printf("    Finished\n");

    if (arguments->verbose) {
        printf ("    Displaying first %i results:\n", VERBOSE_DEPTH);
        result * r = results;
        for( i = 0; i < VERBOSE_DEPTH; i ++ ) {
            printf(  "        point[%i]\n",i );
            for( j = 0; j < RESULT_FIELDS; j ++ ) {
                printf(  "            result[%i] %0161X\n", j, (*r).f[j] );
            }
            r ++;
        }
    }
    return(0);
}


int validate(result *resultsSw, result *resultsHw) {
    result *rs = resultsSw;
    result *rh = resultsHw;
    u_64 rfs, rfh;
    int i, j;
    for( i = 0; i < NBR_POINTS; i++, rs++, rh++ ) {
    for( j = 0; j < RESULT_FIELDS; j ++ ) {
        rfs = (*rs).f[j];
        rfh = (*rh).f[j];
        if (rfs != rfh ) {
        printf("    FAILURE: Mismatch between hardware & software "
                "results for point %i\n", i);
        printf("            s/w: %0161X  h/w: %0161X\n", rfs, rfh);
        return(1);
        }
    }
    }
    printf("    Hardware and software results match.\n");
    return(0);
}
```

```c
/**************************************************************************/
/* On with the main program ...                                         */
/**************************************************************************/
int main (int argc, char **argv) {
    struct arguments arguments;
    int fpga_id;
    err_e e;

    /* Parse any command line options and arguments. Store them in */
    /* the arguments structure. */
    arguments.verbose = 0;
    argp_parse (&argp, argc, argv, 0, 0, &arguments);

    /* Open the FPGA device */
    fpga_id = fpga_open ("/dev/ufp0", O_RDWR|O_SYNC, &e);
    if (e != NOERR) {
        printf ("Failed to open FPGA device. Exiting.\n");
        return(1);
    }

    /* Declare & initialize structures which hold input data */
    point *points = malloc( NBR_POINTS * sizeof(point));
    hypercube *hypercubes = malloc( NBR_HYPERCUBES * sizeof(hypercube));
    printf ("\nGenerating sample data for points & hypercubes\n");
    init_data_structures(&arguments, points, hypercubes );

    /* Declare pointers to result data */
    result * resultsSw;
    result * resultsHw;

    /* Perform search in hardware */
    clock_t t1 = clock();
    printf ("\nSearching for point-hypercube intersections "
            "using hardware\n");
    search_hw(&arguments, fpga_id, points, hypercubes, &resultsHw);
    clock_t t2 = clock();
```

```c
/* Perform search in software */
clock_t t3 = clock();
printf ("\nSearching for point-hypercube intersections "
        "using software\n");
search_sw(&arguments, points, hypercubes, &resultsSw);
clock_t t4 = clock();

/* Make sure the hardware and software results match */
printf ("\nValidating results\n");
validate(resultsSw, resultsHw);

/* Analyze results */
printf ("\nStatistics:\n");
double time_hw = (double)( (t2-t1)/(CLOCKS_PER_SEC/1000));
double time_sw = (double)( (t4-t3)/(CLOCKS_PER_SEC/1000));
printf ("     Hardware time: %8d ms\n", (int)time_hw);
printf ("     Software time: %8d ms\n", (int)time_sw);
printf ("     Factor: %4.2f\n", time_sw / time_hw);

/* Close the FPGA device */
fpga_close (fpga_id, &e);
// TODO cleanup memory

return 0;
}
```

APPENDIX B

search.hcc

```
/*****************************************************************
 * Copyright (C) 2005 Celoxica Ltd. All rights reserved.        *
 *****************************************************************
 *                                                              *
 * Project     :    NRL Eval                                    *
 * Date        :    01 Feb 2006                                 *
 * File        :    main.hcc                                    *
 * Author      :    David Gardner                               *
 *                                                              *
 * Description:                                                 *
 *    Point-Hypercube intersection search                       *
 *                                                              *
 * Date          Version   Author  Reason for change            *
 *                                                              *
 * 16 Feb 2006    0.1       DG      Original version             *
 *                                                              *
 *****************************************************************/


#include <cray_xd1.hch>

// Clock Configuration - Allowable frequencies are from 63MHz to 199MHz
#define CLOCK_RATE 140000000

// Useful constants
#define NBR_HYPERCUBES 64
#define NBR_DIMS 4
#define BYTES_PER_DIM 4
#define BUS_WIDTH 8
#define HC_MEM_SIZE (NBR_HYPERCUBES * NBR_DIMS * BYTES_PER_DIM / BUS_WIDTH)
#define MEGA (1024 * 1024)
```

```
// Configure FPGA Address space - 128MB
//
//      Notes regarding address selection
//
//      1. It's probably best to cover the entire 128MB address space.
//         Otherwise, an errant read to an unused memory area could cause
//         a kernel panic.
//
//      2. Addresses are given as "byte-address".  The PSL will drop 3 bits
//         as appropriate to translate byte addresses to quadword addresses.
//
//      3. BASE and TOP addresses are inclusive. Hence, the "-1" on TOP
//         addresses.
//
//      4. The address spaces for all callbacks must not overlap.
//         In particular, if a read is performed in an address space that
//         is overlapping, the results will be unpredictable (possibly
//         kernel panic).
//
//      5. The low 64MB of address space is "Memory Space" and is subject
//         to "write combining". Address spaces that may be adversely
//         affected by write combining should be at or above the 64MB
//         boundary.
//         For more information refer to the "Processor requests" section
//         of the Cray XD1 FPGA Development document.
//


#define TOP            ( 128 * MEGA )

#define CONFIG_TOP                   ( TOP - 1 )
#define CONFIG_BASE  (  64 * MEGA )                       // 0x4000000

#define POINTS_TOP                   ( CONFIG_BASE - 1)
#define POINTS_BASE  (   0 * MEGA )                       // 0x0000000



// Boilerplate code to setup the FPGA clock and reset signals
#ifdef SIMULATE
```

```
set clock = external;
#else
interface port_in (unsigned 1 user_clk) user_clk_if ();
interface port_in (unsigned 1 reset_n) reset_n_if ();
set clock = internal user_clk_if.user_clk with {
                    rate = (CLOCK_RATE / 1000000) };
set reset = internal !reset_n_if.reset_n;
#endif

// Synchronization flags
static signal unsigned 1 newPointNextCycle = 0;
static unsigned 1 resultRdy = 0;


/*****************************************************************
 * Configuration Registers
 *
 * Addr   Function
 * ----   -------------------------------------------------------
 * 0-63   Hypercube Buffers
 *        This is where we manage the list of hypercubes that we're
 *        searching through.  (This is actually a FIFO, not memory
 *        mapped.)
 *   64   Write: The address to processor memory where results are
 *        to be stored
 *
 ****************************************************************/
// pointer to processor memory where we'll store our results
unsigned 40 resultPtr = 0;
unsigned 64 hypercubes[HC_MEM_SIZE];

macro proc Shift(arry, newVal, DEPTH) {
    par {
        arry[DEPTH-1] = newVal;
        par ( i = 0; i < DEPTH-1; i ++ ) {
            arry[i] = arry[i+1];
        }
    }
}
```

```
macro proc WriteConfigData(addr, data, byteMask) {
    if ( addr[7] == 0 ) par {
        // The processor passes us a 64 bit value which contains
        // hypercube information.
        // As we receive each new datum, we'll shift the previous
        // datums through the array
        Shift(hypercubes, data, HC_MEM_SIZE);
    } else {
        resultPtr = data <- 40;
        RTWriteSetAddr(resultPtr); // this will force a flush if needed
    }
}

// Just for debug purposes, let's support reading from the end of our
// hypercube shift register
macro proc ReadConfigData(addr, data) {
    data = hypercubes[0];
}


/****************************************************************
 * Point input macros
 *
 ****************************************************************/
unsigned 16 x, y, z, a;
unsigned 64 searchResult;

macro proc WritePointData(addr, data, byteMask) {
    par {
        a = data[63:48];
        z = data[47:32];
        y = data[31:16];
        x = data[15:0];
        newPointNextCycle = 1;
    }
}
```

```
macro proc ReadSearchResult(addr, data) {
    data = searchResult;
}



/*****************************************************************
 * Search macro
 *
 * This is where the real work gets done.  As new points are
 * written, they are compared against the hypercubes which were
 * previously loaded into the FPGA.
 *****************************************************************/
macro proc RunSearch() {

    // TODO This macro assumes 64 hypercubes with 4 dimensions each...
    // Perhaps it should be rewritten to automatically configure at
    // compile time.
    unsigned 1 dimTest[NBR_HYPERCUBES][NBR_DIMS*2];
    unsigned 1 result[NBR_HYPERCUBES];
    static unsigned 1 rdyFlag[3] = {0,0,0};

    while(1) par {

        // --- Pipeline stage 0  ---
        // Test min & max conditions of each dimension of each
        // hypercube simultaneously
        rdyFlag[0] = newPointNextCycle;
        rdyFlag[1] = rdyFlag[0];
        par( i = 0; i < NBR_HYPERCUBES; i ++ ) {
            dimTest[i][0] = hypercubes[i*2+0][15: 0] <= x;
            dimTest[i][1] = hypercubes[i*2+0][31:16] >= x;
            dimTest[i][2] = hypercubes[i*2+0][47:32] <= y;
            dimTest[i][3] = hypercubes[i*2+0][63:48] >= y;
            dimTest[i][4] = hypercubes[i*2+1][15: 0] <= z;
            dimTest[i][5] = hypercubes[i*2+1][31:16] >= z;
            dimTest[i][6] = hypercubes[i*2+1][47:32] <= a;
            dimTest[i][7] = hypercubes[i*2+1][63:48] >= a;
        }
```

```
        // --- Pipeline stage 1   ---
        // Consolidate the dimTest flags into one result bit per hypercube
        rdyFlag[2] = rdyFlag[1];
        par( i = 0; i < NBR_HYPERCUBES; i ++ ) {
            result[i] = dimTest[i][0] &&
                        dimTest[i][1] &&
                        dimTest[i][2] &&
                        dimTest[i][3] &&
                        dimTest[i][4] &&
                        dimTest[i][5] &&
                        dimTest[i][6] &&
                        dimTest[i][7];
        }


        // --- Pipeline stage 2   ---
        // Reformat the results into a quadword and signal the
        // "RunResultStream" to send that quadword to the processor
        resultRdy = rdyFlag[2]; // signal the downstream logic
        // TODO: make this a macros and give my fingers a rest.
        searchResult =
            result[63] @ result[62] @ result[61] @ result[60] @
            result[59] @ result[58] @ result[57] @ result[56] @
            result[55] @ result[54] @ result[53] @ result[52] @
            result[51] @ result[50] @ result[49] @ result[48] @
            result[47] @ result[46] @ result[45] @ result[44] @
            result[43] @ result[42] @ result[41] @ result[40] @
            result[39] @ result[38] @ result[37] @ result[36] @
            result[35] @ result[34] @ result[33] @ result[32] @
            result[31] @ result[30] @ result[29] @ result[28] @
            result[27] @ result[26] @ result[25] @ result[24] @
            result[23] @ result[22] @ result[21] @ result[20] @
            result[19] @ result[18] @ result[17] @ result[16] @
            result[15] @ result[14] @ result[13] @ result[12] @
            result[11] @ result[10] @ result[ 9] @ result[ 8] @
            result[ 7] @ result[ 6] @ result[ 5] @ result[ 4] @
            result[ 3] @ result[ 2] @ result[ 1] @ result[ 0];
    }
}
```

```
macro proc RunResultStream() {
    while(1) par {
        if ( resultRdy ) par {
            // send the result to the processor
            RTWriteSequential( searchResult );
        } else delay;
    }
}


/*****************************************************************
 * main
 *****************************************************************/
void main() {
    par {
        // Run the Rapid Transport "driver"
        RunRTIf();

        // Run the application specific logic
        RunRTClient( POINTS_BASE, POINTS_TOP, ReadSearchResult,
                     WritePointData );
        RunRTClient( CONFIG_BASE, CONFIG_TOP, ReadConfigData,
                     WriteConfigData );
        RunSearch();
        RunResultStream();
    }
}
```

Bibliography

[1] http://ati.amd.com/technology/streamcomputing/product_firestream_9170.html, 2008.

[2] Bachmann, C.M., Ainsworth, T.L. and Fusina, R.A., "Improved Manifold Coordinate Representations of Large Scale Hyperspectral Scenes," *IEEE Transactions on Geoscience and Remote Sensing*, Volume 44 (10), 2006. DOI: 10.1109/TGRS.2006.881801

[3] Betz, V. , "Placement for General Purpose FPGAs," in *Reconfigurable Computing*, S. Hauck and A. Dehon (eds.), Morgan Kaufmann, 2008.

[4] Bique, S., Anderson, W. and Lanzagorta, M., "Sorting using the Xilinx Virtex-4 Field Programmable Gate Arrays," in PARA 2008 Conference Proceedings, Trondheim, Norway, 2008.

[5] Bique, S., Anderson, W., Lanzagorta, M., and Rosenberg, R., "Sorting Using the Xilinx Virtex-4 Field Programmable Gate Arrays on the Cray XD1," in Cray Users Group Conference Proceedings, Helsinki, Finland, 2008.

[6] Bique, S. and Rosenberg, R., "Fast Generation of High-Quality Pseudorandom Numbers and Permutations Using MPI and OpenMP on the Cray XD1," in CUG 2009 Conference Proceedings, 2009.

[7] Brown, S. and Vranesic, Z., "Fundamentals of Digital Logic with VHDL Design," McGraw Hill, 2005.

[8] "Celoxica DK4.0 SP1 Beta for Cray XD1 FPGA on Linux," Celoxica, 2006.

[9] conveycomputers.com

[10] Chang, M.L., "Device Architecture," in *Reconfigurable Computing*, S. Hauck and A. Dehon (eds.), Morgan Kaufmann, 2008.

[11] Cong, J. and Pan, P. , "Technology Mapping," in *Reconfigurable Computing*, S. Hauck and A. Dehon (eds.), Morgan Kaufmann, 2008.

[12] Cray XD1^{TM} FPGA Development, S-6400-14, Cray, Inc., 2006.

[13] Diaconis, P. and Shahshahani, M., "Generating a random permutation with random transpositions, Probability Theory and Related Fields," Volume 57 (2), 1981. DOI: 10.1007/BF00535487

[14] El-Ghazawi, T., "High Performance Reconfigurable Computing: A HL Programming Perspective," slides of the tutorial presented at the DoD High Performance Computing Modernization Program Users Group Conference, Denver, June 2006.

[15] Fernando, J.A., "Using FPGAs in High Performance Computing," slides of the tutorial presented at the US Naval Research Laboratory, Washington DC, 2006.

[16] GSL Team, "Other random number generators in GNU Scientific Library: Reference Manual" http://www.gnu.org/software/gsl/manual/html_node/index.html, 2007.

[17] Guccione, S. A., "Configuration Bitstream Generation," in *Reconfigurable Computing*, S. Hauck and A. Dehon (eds.), Morgan Kaufmann, 2008.

[18] JáJá, J., "An Introduction to Parallel Algorithms," Addison-Wesley Publishing Company, 1992.

[19] Kilts, S., "Advanced FPGA Design: Architecture, Implementation, and Optimization," Wiley-Interscience, 2007.

[20] Mogill, J., "Cray XD-1 System Specifics," slides of tutorial presented at the Naval Research Laboratory, Washington, D.C., 2007.

[21] Möhl, S., "The Mitrion-C Programming Language," 1.3.0-001, Mitrionics, 2007.

[22] Morris, K., "A Passel of Processors FPGA and Structural ASIC Journal," http://www.fpgajournal.com/articles_2008/20080617_nvidia.htm, 2008.

[23] Pedroni, V.A., "Circuit Design with VHDL," The MIT Press, 2004.

[24] Rajan, S., "Essential VHDL: RTL Synthesis Done Right," F. E. Compton, 1998.

[25] RG, "Handel-C Language Reference Manual," RM-1003-4.4, Celoxica, 2007.

[26] RG, "Introduction to PDK," Celoxica, 2005.

[27] "Running the Mitrion Virtual Processor on XD1," 1.2-003, Mitrionics, 2007.

[28] Singh, S., "Specifying Circuit Layout on FPGAs," in *Reconfigurable Computing*, S. Hauck and A. Dehon (eds.), Morgan Kaufmann, 2008. DOI: 10.1145/508352.508353

[29] Tenenbaum, J.B., and de Silva, V. and Langford, J.C ., "A Global Geometric Framework for Nonlinear Dimensionality Reduction," *Science*, Volume 290, 2000. DOI: 10.1126/science.290.5500.2319

[30] "The Mitrion Host Abstraction Layer API," 1.3.0-001, Mitrionics, 2007.

[31] "The Mitrion Software Development Kit," 1.3.0-002, Mitrionics, 2007.

[32] Underwood, K.D. and Hemmert, K.S., "The Implications of Floating Point for FPGAs," in *Reconfigurable Computing*, S. Hauck and A. Dehon (eds.), Morgan Kaufmann, 2008.

[33] Vahid, F. and Stitt, G., "Hardware/Software Partitioning," in *Reconfigurable Computing*, S. Hauck and A. Dehon (eds.), Morgan Kaufmann, 2008.

[34] Valente, G. and XtremeData, "To accelerate or not to accelerate…," Programmable Logic DesignLine, 2008.
www.pldesignline.com/howto/208403201

[35] "Virtex-II Pro and Virtex-II Pro X Platform FPGAs: Complete Data Sheet," DS083 (v4.7), Xilinx, Inc., 2007.

[36] "Virtex-4 Family Overview," DS112 (v3.0), Xilinx, Inc., 2007.

[37] http://www.intel.com/technology/visual/microarch.htm, 2009.

[38] http://www.lrz-muenchen.de/services/software/parallel/pgas/index.html, 2009.

[39] http://www.mcs.anl.gov/petsc/petsc-2/, 2008.

[40] http://www.nvidia.com/object/tesla_computing_solutions.html, 2009.

[41] http://www.interactivesupercomputing.com/solutions/solutionshpc.php, 2009.

[42] http://www.xilinx.com/company/history.htm

[43] http://fpga.totallyexplained.com/

[44] W. Gropp, E. Lusk, A. Skjellum, "Using MPI—Portable Parallel Programming wiht the Message Passing Interface, 2nd Edition, The MIT Press, Cambridge, Massachusetts, 1999.

[45] http://www.srccomputers.com

[46] Underwood, K.,"FPGAs vs. CPUs: Trends in Peak Floating-Point Performance, FPGA'04, Monterey, CA, ACM, February 22-24, 2004

[47] Hillis, W.D., Steele, Jr., G.L., Data Parallel Algorithms, Communications of the ACM, v. 29, n12, p1170-1183, Dec. 1986

[48] http://www.dsplogic.com/home

[49] http://www.impulseaccelerated.com

[50] http://www.nallatech.com/index.php/Development-Tools/dime-c.html

[51] http://www.informationweek.com/shared/printableArticle.jhtml; jsessionid=UGFQBENN3GQ41QE1GHRSKHWATMY32JVN?articleID=190400264

[52] Gropp, W.D. "Software for Petascale Computing," Computing in Science and Engineering, v. 11, #5, pp.17–21.

[53] Brodtkorb, A., Dyken, C., Hagen, T., Hjelmervik, J. and Storaasli, O., "State of the Art in Heterogeneous Computing," Scientific Programming, IOS Press, Amsterdam, Netherlands, 2009.

[54] www.chrec.org/CHRECtools.pdf

[55] Holland, B., Nagarajan, K., Jacobs, A. and George, A.D., "RAT: A Methodology for Predicting Perfromance in Application Design Migration to FPGAs," HPRCTA'07, Reno, Nevada, 2007. DOI: 10.1145/1328554.1328560

[56] Marquardt, A., Betz, V., and Rose, J., "Timing-Driven Placement for FPGAs," ACM/SIGDA International Symposium on Field Programmable Gate Arrays, Monterey, CA, February 2000, pp. 203 - 213. DOI: 10.1145/329166.329208

[57] Selvakkumaran, N., Ranjan, A., Raje, S., Karypis, G. "Multi-resouce aware partitioning algorithms for FPGAs with heterogeneous resources," February 2004 FPGA '04: Proceedings of the 2004 ACM/SIGDA 12th international symposium on Field programmable gate arrays, Monterrey, CA, 2004.

[58] Storaasli, O. and Strenski, D.G. Exploring Accelerating Science Applications with FPGAs, NCSA/RSSI Proceedings, Urbana, IL, July 20 , 2007

[59] Storaasli, O. and Strenski, D.G., *Cray XD1 100X Speedup/FPGA Exceeded: Timing Analysis Yields Further Gains*, Proceedings of Cray Users Group (CUG'09), Atlanta GA, May 4-7, 2009.

[60] McGill, J., formerly of Mitrionics, private communication.

[61] Morris, K., "A Passel of Processors," FPGA and Structured ASIC Journal, June 18, 2008.

Authors' Biographies

STEPHEN BIQUE

Dr. Stephen Bique is a computer scientist in the Center for Computational Science at the US Naval Research Laboratory in Washington, DC. He received a PhD in Computer Science from the University of Joensuu in Finland. Previously, Dr. Bique was Associate Professor at Virginia State University, Assistant Professor at the University of Alaska Fairbanks, and he held a comparable position at the University of Kuopio in Finland. He has also worked as software developer and consultant for PACT Corporation in Munich, Germany on the "XP128" reconfigurable system-on-a-chip.

MARCO LANZAGORTA

Dr. Marco Lanzagorta is Technical Fellow and Director of the Quantum Technologies Group of ITT Corporation. In addition, Dr. Lanzagorta is an Affiliate Associate Professor at George Mason University and co-editor of the Synthesis Lectures on Quantum Computing published by Morgan & Claypool. Dr. Lanzagorta received a PhD in theoretical physics from Oxford University and in the past he worked at the US Naval Research Laboratory, the European Organization for Nuclear Research (CERN) in Switzerland, and the International Center for Theoretical Physics (ICTP) in Italy.

ROBERT ROSENBERG

Dr. Robert Rosenberg is an Information Technology Specialist in the Center for Computational Science at the US Naval Research Laboratory in Washington, DC. He received a PhD in theoretical physical chemistry from Columbia University. In the past he served on both the DoD High Performance Computing Modernization Program Requirements and Benchmarking teams. His past work at NRL includes both scientific programming and scientific visualization.

Printed in the United States
by Baker & Taylor Publisher Services